Nonlinear Effects in Model Lattices of Metals

Solitons, Discrete Breathers, Quasi-Breathers, Shock Waves

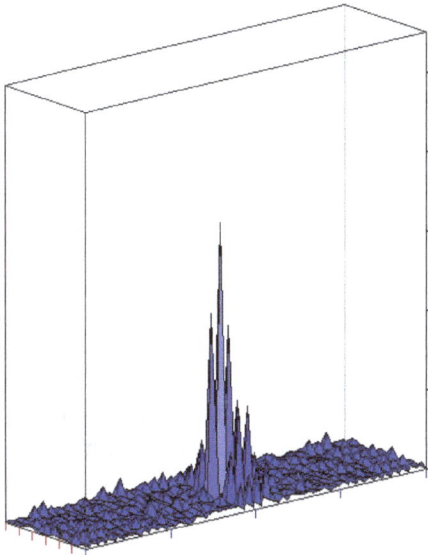

Mikhail D. Starostenkov
Pavel V. Zakharov
Artem V. Markidonov
Pavel Y. Tabakov

Published by **Materials Research Forum LLC**
Millersville, PA 17551, USA

Published as part of the book series
Materials Research Foundations
Volume 156 (2024)
ISSN 2471-8890 (Print)
ISSN 2471-8904 (Online)

Print ISBN 978-1-64490-288-2
ePDF ISBN 978-1-64490-289-9

Distributed worldwide by

Materials Research Forum LLC
105 Springdale Lane
Millersville, PA 17551
USA
https://www.mrforum.com

Printed in the United States of America
10 9 8 7 6 5 4 3 2 1

Table of Contents

Preface

This book is devoted to an overview of several nonlinear effects arising in discrete crystal lattices with realistic potentials. The main attention is paid to discrete breathers, quasi-breathers, soliton waves, and shock waves. All studies considered in the paper are performed with the help of atomistic simulation methods. The approach used is in good agreement with the available theoretical and experimental data. The first chapter is devoted to a discussion of nonlinear processes and approaches in atomistic modeling. Most of the calculations were carried out employing the molecular dynamics method. The next three chapters deal with discrete breathers in different crystals and conditions. Emphasis is placed on the fact that such objects are quasi-periodic, thus amenable to statistical evaluation. Chapter five discusses the mechanism of nonlinear supratransmission and the role of breathers in the nucleation of soliton waves, which can propagate over substantial distances in discrete periodic structures. The final sixth chapter discusses the effect of shock waves on the defect structure in fcc crystals. The conclusion reflects the main scientific results and further prospects for the development of this direction.

Keywords

Soliton, Discrete Breather, Quasi-Breather, Shock Waves, Supersonic Waves, Molecular Dynamics, Self-Organization, Nonlinearity, Metals

Acronyms

ADB	Antisymmetric discrete breather
DFT	Density functional theory
DB	Discrete breather
DOS	Densities of states
DXA	Dislocation extraction algorithm
EAM	Embedded atom method
LAMMPS	Large-scale Atomic/Molecular Massively Parallel Simulator
PFs	Frenkel pairs
SFT	Stacking fault tetrahedra

Introduction

The nonlinearity of interatomic bonds in many-particle systems leads to a wide range of various phenomena and processes that have a significant impact on the structural and energy transformations of materials. The importance of these often "subtle" manifestations should not be underestimated because they manifest themselves in critical situations and systems far from equilibrium. It is this factor that attracts many researchers to such phenomena as solitons, discrete breathers, crowdions, vortices, chimeras, shock waves, and many others.

Discovered in the second half of the 20th century, such effects of energy transport from the surface deep into the crystal as the effect of low doses and the long-range effect still do not have an unambiguous interpretation. Some authors suggest that it is the soliton mechanism that is one of the main ones in the ion irradiation of crystals. At present, there are several concepts regarding the mechanisms that cause the directed drift of crystal lattice defects under external influences that generate Frenkel pairs (PFs). According to one of them, the movement of defects is caused by elastic waves resulting from the recombination of unstable Frenkel pairs. In the other one, the soliton mechanism of defect transfer is considered.

At the same time, new research works appear, where the authors tend to consider the excitation of the atomic system of a crystal under external intense influences from the standpoint of various types of solitons. For example, the annealing of defects in a germanium crystal at a great distance from the surface, while the crystal surface was exposed to a low-energy plasma discharge, this effect was explained from the standpoint of discrete breathers. Active attempts are also being made to detect localized modes in metals and alloys by experimental methods, but there are many difficulties, in particular, there is the problem of excitation of such objects, as well as the need to distinguish them from other types of lattice vibrations. It is worth mentioning such a type of solitons as crowdions. The static and dynamic properties of crowdions differ significantly from the properties of localized interstitial atoms and vacancies; therefore, they are considered in crystal physics as an independent type of intrinsic crystal lattice defects. For example, the dynamics of a crowdion differs from that characteristic of point defects, its mobility is very high along the close packing direction (even at low temperatures, when quantum effects appear) and is zero for all other directions. At the moment, such objects as N-crowdions are considered, and there is a significant amount of research on crowdions and their contribution to the properties of crystals.

It should be noted that the increase in the microhardness of samples at depths exceeding the calculated ion ranges is of great interest to researchers. One of the reasons for the increase in microhardness, in this case, is the formation of interstitial dislocation loops due to the coagulation of point defects that arise during the passage of shock waves through the volume of the sample. Recent studies show the role of shock waves in such phenomena as the superplasticity of crystals.

The lack of a clear understanding of the described effects is a motivating factor in the study of such processes at the atomic level and the possible contribution of the soliton mechanism to the energy transport through the crystal.

A change in the structure of material under intense external influence (radiation) is a complex, multilevel phenomenon with a wide temporal and spatial range. Due to the complexity of studying such processes at the atomic level, it seems most appropriate to use the computer simulation method, which is currently the same recognized method of cognition as the experimental and theoretical methods.

With the help of a computer model, it is possible to test theoretical developments, explain and predict phenomena that have not yet been fully elucidated by other research methods, and in addition, computer simulation is favorably distinguished by the relative cheapness of obtaining data. The method of molecular dynamics was chosen as the main method of computer simulation in our works because it allows us to conduct experiments with given velocities of atoms and compare the dynamics of the processes under study in real-time. It allows for solving problems related to the problems of structural energy transformations, both in crystalline and non-crystalline materials. In addition, this method makes it possible to calculate many system properties, both thermodynamic (for example, energy, pressure, entropy) and kinetic (diffusion coefficients, atomic vibration frequencies). Also, the use of computer simulation allows us to identify fundamental patterns at the lowest cost.

The study is carried out using the MD modeling package: LAMMPS Molecular Dynamics Simulator, XMD (Molecular Dynamics for Metals and Ceramics), as well as proprietary software. The main task in modeling the atomic structure is the description of the forces of interatomic interaction since the correctness of further modeling largely depends on this description. The simplest way to describe interatomic interaction is to use pair potential functions. But, strictly speaking, the representation of energy as a sum of pair interactions is valid only for crystals of inert gases, in which interatomic bonds are determined by van der Waals forces. In metals, between atoms, due to the effects of electronic distribution, in addition to direct ion-ion interaction, there is also indirect interaction. Besides, the pair potentials incorrectly describe the elastic properties of crystals, overestimate the formation of a vacancy in a metal, do not describe the decrease in the bond energy per bond with an increase in the coordination number, etc. Methods for describing interatomic interactions based on the density functional theory do not have the disadvantages described above. Here we use both the pair potentials, which allow us to obtain qualitative regularities and the Johnson potential calculated using the embedded atom method (EAM-potential), as well as other EAM potentials, which will ensure greater reliability of the results. Calculations by the MD method should be verified by a more accurate density functional method (DFT), which describes the dynamics of electrons in outer shells and, thus, allows us to naturally take into account the change in the polarization of atoms during oscillations. However, this method has some limitations associated with the complexity of the calculations. The experience of using this method shows that even the use of a supercomputer does not provide a high speed of calculations.

However, some problems that do not require a large volume of atoms can be solved using this method. Therefore, to refine the results obtained in the framework of the classical method of molecular dynamics, the density functional theory will be applied to some of the problems under consideration.

Thus, this book is intended to present an overview of some of the nonlinear effects that arise in crystal lattices under intense external influences. Their consideration will be carried out utilizing atomistic modeling methods using realistic interaction potentials.

Chapter 1. Nonlinear Dynamics and Modeling

1.1 Soliton-type waves and supersonic waves in crystals

Solitons are stabilized by the presence of integrals of motion such as total energy or momentum. Such solitons are called dynamic and exist as stationary states only to the extent that energy and momentum are conserved. If arbitrarily small perturbations destroying these integrals of motion are included in the equations of motion, then dynamic solitons can be eliminated.

At the same time, it is extremely difficult to talk about the stationarity of such objects in real systems, due to the presence of all kinds of perturbations in real physical media and simulated systems.

An example of a dynamic soliton can be precise discrete breathers, i.e., space-localized and time-periodic high-amplitude excitations in nonlinear discrete structures with translational symmetry. They are also called internal localized modes or nonlinear localized excitations.

A discrete breather, as an object strictly periodic in time, is obtained in numerical simulation only in the case of an ideal adjustment of the initial conditions of the Cauchy problem to a certain low-dimensional manifold in a multidimensional space of all possible initial values of the coordinates of individual particles and their velocities. Such fine-tuning is difficult to implement even in a computational experiment. Moreover, it is practically impossible to do this when setting up any physical experiments, especially in cases where breather-like objects arise spontaneously.

In this regard, the concept of quasi-breathers was put forward, as some dynamic objects localized in space, but not strictly periodic in time. At the same time, a certain criterion for the proximity of a quasi-breather to the corresponding exact breather was formulated, based on the calculation of the root-mean-square deviation $\eta(t_k)$ of the oscillation frequencies of individual breather particles found on a certain interval in the vicinity of the moment t_k, and the calculation of the root-mean-square deviation of the oscillation frequencies of the selected j-th breather particles at different time intervals.

Unlike exact discrete breathers, quasi-breathers are not strictly time-periodic dynamic objects, although they are localized in space. They arise for any sufficiently small deviations from the exact breather solutions in the multidimensional space of all possible initial conditions when solving the Cauchy problem for the original differential equations, since in this case there is no complete suppression of the contributions from oscillations of peripheral particles with their own frequencies. Thus, the "weakening of the dictatorship" on the part of the breather core leads to the presence in the breather solution of small contributions having different frequencies. In the case of the considered symmetric breather, the core is also formed by one central particle. These small contributions can be detected in the oscillations of all particles in the chain, in particular, central ones. If one finds sufficiently accurately the frequencies of oscillations of all particles of the quasi-breather, calculated on a certain time interval near $t = t_k$, then they will not be strictly the same. In light of this, it makes sense to find the standard deviations $\eta(t_k)$ of the oscillation frequency of various particles of the breather from the average breather frequency ϖ.

The larger the value $\eta(t_k)$, the more the quasi-breather solution differs from the exact breather solution, for which $\eta(t_k) = 0$ at any time t_k. The standard deviation gives an absolute estimate of the spread measure. Therefore, to understand how large the spread is relative to the values themselves (i.e., regardless of their scale), a relative indicator is required. This indicator is called the coefficient of variation.

The concept of localization of vibrational energy arising due to anharmonicity in nonlinear models of ideal crystal lattices of various dimensions has experienced intensive development over a quarter of a century since the appearance of the first publication. During this time, the possibility of the existence of non-linear localized objects has been proven rigorously using theorems and a set of results obtained by numerical integration of equations that describe the dynamics of various models and do not have analytical solutions. At the same time, when a solid body is irradiated with a beam of accelerated ions, some of them are reflected from the surface, while the rest penetrates deep into the volume of the material, slowing down in it. In this case, the kinetic energy of the ions is wasted during elastic collisions with the nuclei of atoms of the material and the excitation of the electronic subsystem. As a result of elastic collisions, the atoms of the body can be knocked out of their equilibrium positions, and elastically colliding with other atoms, create a stream of knocked-out atoms, forming a cascade of atomic collisions. The atoms in the cascade expend energy on the formation of point defects. In addition, the resulting strong nonequilibrium temperature fluctuations lead to the formation of so-called post-cascade shock waves. Their occurrence is due to the difference between the thermalization time of atomic oscillations in a certain finite region and the time of heat removal from it. As a result of a sharp expansion of a strongly heated region, an almost spherical shock wave is formed. The emergence of nanosized areas of energy-explosive release that generate shock waves, is a common phenomenon for any type of corpuscular irradiation. However, this fact is practically not taken into account when studying the behavior of condensed matter under conditions of radiation exposure. Therefore, the discovery of new phenomena and processes initiated by post-cascade shock waves and their orientation towards the formation of unique modified atomic structures is a promising and topical area of radiation materials science. The effects associated with the propagation and generation of post-cascade shock waves are called radiation-dynamic. In this case, high-speed cooperative atomic displacements are carried out, which are a process that proceeds at supersonic speed. Revealing the mechanisms of the effect of high-speed cooperative atomic displacements, which can be considered as shock and supersonic waves, on structural changes occurring in metals with an fcc lattice is also relevant for understanding the processes occurring at the atomic level in crystals.

It should be noted that the increase in the microhardness of samples at depths exceeding the calculated ion ranges is of great interest to researchers. One of the reasons for the increase in microhardness, in this case, is the formation of interstitial dislocation loops due to the coagulation of point defects that arise during the passage of shock waves through the volume of the sample. Recent works show the role of shock waves in such phenomena as the superplasticity of crystals.

Thus, in this book, the focus will be on discrete breathers, quasi-breathers in various crystals, solitary, and shock waves.

1.2 Method of molecular dynamics and first-principles calculations in the study of soliton-type waves

The method of molecular dynamics using classical mechanics is based on solving the system of Newton's ordinary differential equations of motion for a system of atoms. This method, in comparison with other methods of computer simulation, has several important advantages. It allows for solving problems related to the problems of structural and energy transformations in both crystalline and non-crystalline materials. In addition, this method allows us to calculate any properties of the system - both thermodynamic (for example, energy, pressure, entropy) and kinetic (diffusion coefficients, atomic vibration frequencies). Moreover, with this method, it is possible to measure the dynamics of the studied processes in real-time.

The main task is to choose the potential of interatomic interaction. When studying non-linear localized states in crystals, we used both Mosre pair potentials and potentials obtained by the embedded atom method - EAM-potential.

Let us consider in more detail each of the types of potentials.
In the case when the interatomic interaction was given by the Morse pair central potential, which has the form:

$$\phi(r_{ij}) = D\beta e^{-\alpha r_{ij}}(e^{-\alpha r_{ij}} - 2), \tag{1}$$

where D is the energy parameter corresponding to the depth of the potential well, α is the parameter that determines the rigidity of interatomic bonds, $\beta = e^{\alpha \cdot r_0}$, r_0 is some averaged equilibrium distance over the coordination spheres, in which the interaction between atoms is taken into account.

The method for determining the parameters of the Morse potential was first proposed by Girifalco and Weiser. The force acting on the i-th atom from the j-th is equal to:

$$\vec{F}(r_{ij}) = -2D\alpha \left[\left(\beta e^{-\alpha r_{ij}} - \frac{1}{2} \right)^2 - \frac{1}{4} \right]. \tag{2}$$

The Morse potential includes two components, one of which is a hard exponential repulsion, and the other is a softer exponential attraction. Therefore, this potential can be used to describe a stable close-packed lattice.

Parameters D, β, and α were determined from the following conditions:

$$\frac{1}{2}\sum_{i=1}^{z}\eta_i\,\phi_{V=V_0} = E_S, \frac{1}{2}\sum_{i=1}^{z}\eta_i\left(\frac{\partial\phi}{\partial V}\right)_{V=V_0} = 0, -V_0\cdot\left(\frac{\partial P_S}{\partial V}\right) = K_0. \tag{3}$$

Here E_S is the energy of sublimation of crystal atoms; K_0 is the bulk modulus of elasticity; P_s is the pressure of isentropic compression; V_0 and V are the specific volumes in the initial and deformed states; η_i is the number of atoms in the i-th coordination sphere. When calculating the parameters, ten coordination spheres were taken.

Strictly speaking, the representation of energy as a sum of pair interactions is valid only for crystals of inert gases, in which interatomic bonds are determined by van der Waals forces. In metals, between atoms, due to the effects of electronic distribution, in addition to direct ion-ion interaction, there is also indirect interaction. In addition, pair potentials incorrectly describe the elastic properties of crystals, overestimate the formation of a vacancy in a metal, do not describe the decrease in the bond energy per bond with an increase in the coordination number, etc. Methods for describing interatomic interactions based on the density functional theory do not have the disadvantages described above. Therefore, in our work, we also used the Johnson potential calculated using the embedded atom method.

In the simulation, the potential energy of the i-th atom is defined as follows:

$$U_i = F_\alpha\left(\sum_{i\neq j}\rho_\alpha\beta\left(r_{ij}\right)\right) + \frac{1}{2}\sum_{i\neq j}\phi_\alpha\beta\left(r_{ij}\right) \tag{4}$$

where r_{ij} is the distance between the i-th and j-th atoms, $\phi_\alpha\beta$ is the function of the pair potential, ρ_α is the contribution to the charge density of electrons from the j-th atom at the location of the i-th atom, and F is the " immersion" function which represents the energy required to place the i-th atom of type α in the electron cloud.

The EAM method is a multiparticle potential and, since the electron cloud density is the sum of the contributions from a large number of atoms, in practice, to reduce the complexity and, accordingly, the calculation time, the number of neighbors is often limited by the so-called "cutoff radius".

To apply the method to simple one-component systems of atoms, three scalar functions must be specified: the immersion function, the pair interaction function, and the electron cloud density distribution function. For binary alloys, already seven functions are needed: three pair interaction functions (A-A, B-B, A-B), two immersion functions and two electron cloud density distribution functions. Usually, these functions are available in tabular form and are interpolated by cubic splines.

While good quality EAM potentials exist for a significant number of elemental metals, the potentials for alloy systems are few and typically require significant development effort, making rapid uptake of new alloy compositions quite challenging. Interatomic potentials describing alloys are often created by combining several previously developed elemental potentials and fitting functions to experimental data or from *ab initio* data for alloy systems (i.e. mixing enthalpies and moduli). Along with adjusting the cross-interaction conditions, it is possible to improve the quality of the potential by making changes to the functions that affect the calculated energies of the alloy rather than the energies of the pure elements. These changes in elemental components are unique, which means that they must be adjusted when additional elements are added to the potential. This makes the development of multicomponent potential alloys slow and limits accuracy as more components are added (the number of fitting parameters becomes impractical for optimization in systems with more than four or five components).

The Johnson alloy model was proposed to create multi-component EAM potentials without significant computational cost. The main idea of this method is to create new functions for a pair of heterogeneous types (cross-potentials), which are necessary to create the potential of the alloy as the electron density of the weighted average elemental components based on an empirical model:

$$\varphi_{\alpha\beta}(r) = \frac{1}{2}\left(\frac{f_{\beta}(r)}{f_{\alpha}(r)}\varphi_{\alpha\alpha}(r) + \frac{f_{\alpha}(r)}{f_{\beta}(r)}\varphi_{\beta\beta}(r)\right). \tag{5}$$

It is important to note that this method uses a different version of EAM, in which the electron density function (f_{α}) is specific to a single atom type only, and not to a pair of types, as in the Finnis-Sinclair EAM potential. In order to provide the best possible alloy potentials between different elements, this method only needs to find compatible values for the electron density functions of all elements from the database and ensure that the embedding functions of each potential are zero at the equilibrium electron density for that species. Consistent values are set by introducing a scaling parameter for each electron density function to leave the elemental input energy unchanged.

Despite the obvious advantages of embedded atom potentials over pair potentials, they also have limitations due to the assumptions described. Therefore, for better calculations, *ab initio* modeling based on the solution of the Schrödinger equation is used.

Non-empirical methods are closest to reality, but calculations using them require large computing power and many hours of computer time. In the literature, these methods are called *ab initio* and they are based on the solution of the Schrödinger equation. In principle, both electrons and nuclei should be taken into account in these methods, but, as a rule, the Born-Oppenheimer approximation is used, which does not take into account the motion of nuclei and assumes that electrons move in the potential created by a system of fixed nuclei, and nuclear motion is studied already based on experience. Thus, *ab initio* calculation methods, in contrast to the methods of molecular mechanics and semiempirical methods, allow taking into account such parameters as the Coulomb interaction of electrons with nuclei and between themselves, the electrostatic interaction of nuclei and, if necessary, nonrelativistic effects.

In the case of crystals, the Born-Oppenheimer approximation is generally accepted, which makes it possible to consider separately the fast motion of electrons and the slow motion of nuclei. Light electrons quickly adjust to the instantaneous configuration of the ionic subsystem, creating an external field in which heavy nuclei move. Schrödinger equation (SE) for a system of electrons can be written as

$$i\hbar \frac{\partial \Psi(r_1, s_1, ..., r_N, s_N, t)}{\partial t} = \hat{H} \Psi(r_1, s_1, ..., r_N, s_N, t) \tag{6}$$

where r_i and s_i are variables corresponding to the spatial and spin coordinates of electrons. In the case when the Hamilton operator $\hat{H}$ does not explicitly depend on time, the problem is reduced to finding its eigenfunctions and eigenvalues:

$$\hat{H} \Psi(r_1, s_1, ..., r_N, s_N) = E \Psi(r_1, s_1, ..., r_N, s_N) \tag{7}$$

The Hamilton operator can be represented as: $\hat{H} = \hat{T} + \hat{U} + \hat{V}$, where

$\hat{T}$ is the operator of the kinetic energy of the system N_e of electrons:

$$\hat{T} = -\frac{\hbar^2}{2m_e} \sum_{j=1}^{N_e} \nabla_j^2 \tag{8}$$

$\hat{U}$ is the operator of the interaction of electrons with each other:

$$\hat{U} = -\frac{1}{2} \sum_{j \neq k=1}^{N_e} \frac{q_e^2}{|r_j - r_k|} \tag{9}$$

$\hat{V}$ is the operator of the interaction of electrons with other parts of the system, represented as an external potential of the subsystem of nuclei with atomic numbers Z_k:

$$\hat{V} = \sum_{k=1}^{N_{nuclei}} \sum_{j=1}^{N_e} \frac{Z_k q_e^2}{|r_j - R_k|}$$

The approach based on the Schrödinger equation (6) assumes the expression of all physical quantities in terms of the many-particle wave function of the problem, which is the solution of the Schrödinger equation (SE). It is well known that an analytical solution of the Schrödinger equation can only be obtained for simple model systems. One of the most common methods for the approximate solution of SE is the self-consistent Hartree-Fock (HF) field method [71, 72]. In this method, the many-particle wave function is written as a Slater determinant, composed of single-particle wave functions $\psi_j(x_i)$:

$$\Psi(r_1, s_1, ..., r_N, s_N) = \frac{1}{\sqrt{N_e!}} det[\psi_j(x_i)]$$

(10)

where i and j number respectively the rows and columns of the matrix composed of the functions $\psi_j(x_i)$, and x_i includes the spatial and spin parts $x_i=(r_i, s_i)$. In itself, the assumption about the form of the wave function (10) is a rather rough approximation. As a result of substitution (10) into stationary SE (8), the latter is split into N_e-coupled integrodifferential equations for one-electron functions $\psi_j(x_i)$. In the Hartree-Fock method, the interaction between electrons is described within the framework of a self-consistent field, which includes both Hartree terms (they have functions of the type $\psi_j{}^* \cdot \psi_j$ under the integral sign) and exchange-correlation terms (functions of the type $\psi_j{}^* \cdot \psi_k$).

At present, many software products allow the implementation of first-principles calculations: Quantum ESPRESSO, Gaussian, CPMD, ABINIT, VASP, CRYSTAL, and others.

In this work, the Quantum ESPRESSO program was used, which is a complex set of open-source programs for calculating the electronic structure and modeling materials at the nanoscale. It is based on density functional theory, plane waves, and pseudopotentials.

Due to the open-source code, Quantum ESPRESSO is currently a set of independent and often inconsistent modifications. The most common is the basic assembly of the program, which contains all the components necessary for calculations and post-processing, as well as some plug-ins to perform complex tasks. Separately, it is worth noting that Quantum ESPRESSO supports most methods of parallelizing calculations, including GPU acceleration, which seems to be the most promising for researchers who do not have access to super-powerful computers.

The standard Quantum ESPRESSO package allows us to solve a wide range of computational problems. With the help of the programs included in it, it is possible to carry out a self-consistent calculation of the total energy of the system, the forces acting on atoms, and interatomic bonds. The basic set also includes programs for modeling molecular dynamics from first principles, there are two types of dynamic calculations: Car-Parrinello and Born-Oppenheimer. Structural optimization is possible. Quantum ESPRESSO can be used to calculate such parameters as an electron density distribution, phonon density of states, and dielectric tensors. The above examples of using this package show a far from complete list of program features.

Most tasks require visualization of the result, for this Quantum ESPRESSO has some programs for post-processing the received data. With preliminary calculations, these programs make it possible to obtain the distribution of electron density, the density of states, dispersion curves, etc. Compatibility with third-party visualization programs is also provided.

For all calculations, Quantum ESPRESSO uses periodic boundary conditions so that the result is close to reality, the program provides support for all types of Bravais lattice, and the ability to set your own symmetry and translation vectors.

Numerical modeling based on molecular dynamics has a drawback associated with the fact that the interaction of the elements of the system cannot be described otherwise than with the help of various kinds of phenomenological potentials. An alternative is *ab initio* modeling based on the principles of quantum mechanics, the basis for this theory is the Schrödinger equation. Since the many-particle Schrödinger equation cannot be solved exactly, in the quantum mechanical description of atoms, molecules, crystals, and various nanoscale physical objects, one has to resort to a number of approximations.

The main array of calculations was performed using the classical method of molecular dynamics using the Morse pair and EAM potentials. *Ab initio* calculations require a large number of machine-hours, as a result of which calculations of some parameters for the Pt_3Al crystal were carried out.

Chapter 2. Discrete breathers in a Pt₃Al crystal

2.1 Discrete breathers with soft nonlinearity in 3D and 2D Pt₃Al models with Morse pair potential

Let us consider the conditions for the existence of discrete breathers in a three-dimensional model of an A_3B crystal with Morse potential parameters for Pt₃Al. The potential parameters are given in Table 1. The total number of atoms in the cell was 8400. The initial temperature was set equal to 0 K.

Table 1: *Morse potential parameters for Pt₃Al.*

	Al-Al	Pt-Pt	Pt-Al
α, Å^{-1}	1.02658	1.5820	1.3501
β	27.4979	102.89	63.124
D, eV	0.318	0.71	0.5048

To excite a discrete breather, one of the Al atoms was removed from the equilibrium position by imposing a displacement to it in a certain direction, after which the computational cell was relaxed.

A series of experiments established that discrete breathers can exist only in strictly defined directions. For the fcc structure under consideration, these directions are shown by arrows in Figs. 1. Note that they correspond to the direction to the nearest Al atoms.

When an attempt was made to excite the discrete breathers by deflecting the Al atom, either damping of the oscillations occurred in other directions or their polarization in the indicated directions. For example, the deviation of an Al atom along the diagonal of a cubic cell towards another Al atom led to the damping of oscillations within 1–3 ps.

Al atoms, whose high-amplitude vibrations are polarized in the above directions, can localize the energy transferred to them. Fig. 2 shows the temperatures for Al and Pt atoms as a function of time. It can be seen that the energy given to the Al atom over time is not transferred to Pt atoms and is not dissipated.

The maximum amplitude of the discrete breather (DB) oscillations, which was achieved in Pt₃Al model cells, was 0.98 angstroms. In this case, the Al atom was initially given a deviation from the equilibrium position of 1.2 angstroms. After primary relaxation, part of the energy was transferred to neighboring atoms, which performed vibrations consistent with the vibrations of the Al atom.

The stroboscopic pattern of oscillations for the 3D case is shown in Fig. 3. This figure makes it possible to estimate the degree of spatial localization of a discrete breather with a soft type of nonlinearity. And also to see the involvement of neighboring atoms in coordinated vibrations.

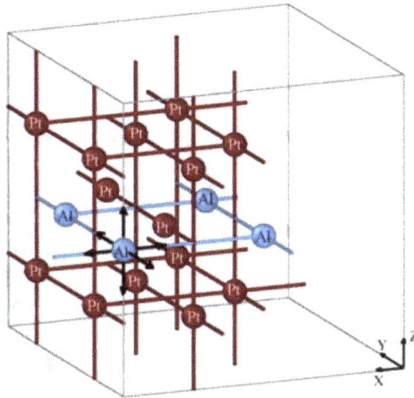

Figure 1: Directions of vibrations of atoms carrying a nonlinear localized mode in a three-dimensional Pt_3Al cell.

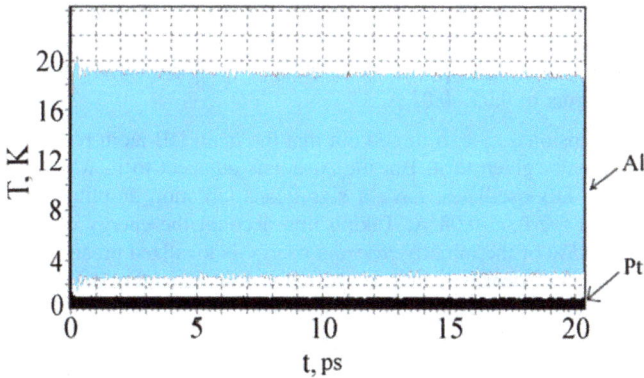

Figure 2: Temperature curves for an excited DB in a three-dimensional Pt_3Al cell.

Let us emphasize the differences between the two- and three-dimensional models in the transfer of energy to neighboring atoms during the first few vibrations. They consist in the fact that in the three-dimensional case, more energy is transferred to the neighboring atoms of the light sublattice in the direction of the DB vibrations. This can be explained by the polarization of atomic vibrations in the 3D model in the direction of the light component of the alloy. In a 2D model, this is not possible, since a light atom is surrounded by heavy ones.

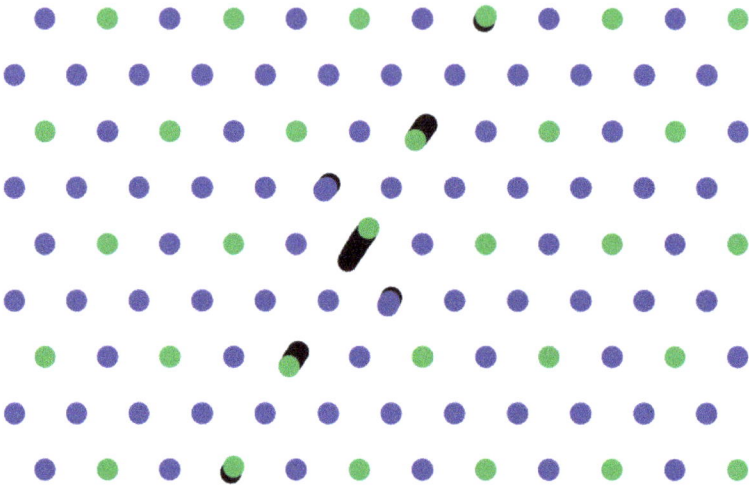

Figure 3: *Stroboscopic pattern of the discrete breather, the (111) plane is visualized.*

Calculations show that in the 2D case, 20% of the energy initially given to the Al atom is redistributed mainly to neighboring Al atoms located in the second coordination sphere within 0.2 ps. At the same time, their oscillation amplitude is an order of magnitude smaller than the oscillation amplitude of an atom carrying a localized mode and amounts to values of the order of 0.05 - 0.01 Å.

In the three-dimensional case, it turned out that the main DB atom retains about 50% of the energy originally given to it. But the Al atoms adjacent to it, which are in the row along which the DB oscillates, have a significant vibration amplitude of 0.3-0.45 Å, while all the rest are 0.05-0.08 Å. Taking into account the energy of these atoms, we obtain that 80 - 85% of the initially reported energy is localized precisely on them in the 3D case (Fig. 3). The linear frequency of the main atom, as well as neighboring Al atoms, was 11.9 THz, with an initial amplitude of 0.72 Å.

To obtain the dependence of the frequency on the vibration amplitude, we preliminarily obtained the density of the phonon states of the crystal under consideration (Fig. 4). The phonon spectrum was calculated in two ways. In the first case, Fig. 4 and the parameters were taken directly from the model, using the software developed by us. The second method consisted in the theoretical calculation of the density of phonon states, according to the following scheme below.

Let us denote the atoms of the crystal by four indices m, n, l, k, where $\infty < m, n, l < \infty$ determine the number of the primitive cell of the infinite crystal, and $l \leq k \leq K$ the number of the atom within one primitive cell. Then the radius vector of an arbitrary superstructure atom will have the form:

$$r = mw_1 + nw_2 + lw_3 + v_k, \tag{11}$$

The equations of motion of an atom with indices m, n, l, k have the form:

$$M_k \ddot{U}_{m,n,l,k} = \sum_{i,j,\xi,\eta} F_{i,j,\xi,\eta}, \tag{12}$$

where $F_{i,j,\xi,\eta}$ is the force acting on the atom m, n, l, k from the side of the atom with indices i, j, ξ, η, and the summation is performed over all neighbors of the atom m, n, l, k within the potential cutoff radius.

Let us find the exact solutions of the linearized equations of motion in the form of a linear superposition of low-amplitude waves of the form:

$$\begin{Bmatrix} u_{x,m,n,l,1} \\ u_{y,m,n,l,1} \\ \dots \\ u_{y,m,n,l,K} \\ u_{z,m,n,l,K} \end{Bmatrix} = U_0 \begin{Bmatrix} U_{x1} \\ U_{y1} \\ \dots \\ U_{yK} \\ U_{zK} \end{Bmatrix} exp(iq_x m + iq_y n + iq_z l - i\omega t), \tag{13}$$

where, $u_{x,m,n,l,k}$, $u_{y,m,nl,k}$, $u_{z,m,n,l,k}$ are the components of the displacement vector of atoms of an infinite crystal; $U_{x,k}, U_{y,k}, U_{z,k}$ are the components of the normalized vector U of dimension $3K$, which determines the eigenmode of the phonon oscillations; u_0 is the phonon amplitude; q_x, q_y, q_z are the wave numbers; ω is the frequency; t is the time; i is the imaginary unit.

Substituting (13) into the system of equations of motion of atoms (12), linearized in the vicinity of their equilibrium positions and written for one primitive cell, we obtain an eigenvalue problem of the form:

$$\begin{bmatrix} C_{11} & \dots & C_{1,3K} \\ \dots & \dots & \dots \\ C_{3K,1} & \dots & C_{3K,3K} \end{bmatrix} \begin{Bmatrix} U_{x1} \\ U_{y1} \\ \dots \\ U_{yK} \\ U_{zK} \end{Bmatrix} = -\omega^2 \begin{Bmatrix} M_1 U_{x1} \\ M_1 U_{y1} \\ \dots \\ M_K U_{yK} \\ M_K U_{zK} \end{Bmatrix}. \tag{14}$$

The eigenvalue problem (14) has a dimension equal to the number of degrees of freedom of atoms that make up one primitive crystal cell, that is, in the three-dimensional case $3K$. Having solved the eigenvalue problem, we obtain $3K$ squares of eigenfrequencies ω^2 and $3K$ eigenvectors. Generally speaking, an eigenvector is complex:

$$U = Ui_{Im_{Re}}, \tag{15}$$

As a solution to the linearized equations of motion, one can use the real

$$U_{Re}\cos\left(q_x m + q_y n + q_z l\text{-}\omega t\right) - U\sin\left(q_x m + q_y n + q_z l\text{-}\omega t\right)_{lm} \tag{16}$$

or imaginary part

$$U_{Im}\cos\left(q_x m + q_y n + q_z l\text{-}\omega t\right) - U_{Re}\sin\left(q_x m + q_y n + q_z l\text{-}\omega t\right) \tag{17}$$

According to these equations, the density of phonon states was calculated and presented in Fig. 4b.

The obtained densities of phonon states agree well with each other, having discrepancies in tenths of a THz in the width of the optical branch of the PS, which does not affect the qualitative result. In further studies, we used the method of obtaining the density of phonon states of a model crystal directly from the model using the Morse potential, which made it possible to consider the change in the geometry of the crystal, for example, during deformation by the phonon spectrum and temperature.

As already noted, one of the main characteristics of the DB is the dependence of its frequency on the oscillation amplitude (Fig. 5). This dependence will make it possible to indirectly judge the DB lifetime. The deeper the breather frequency goes into the band gap of the phonon spectrum the more time is required for energy dissipation.

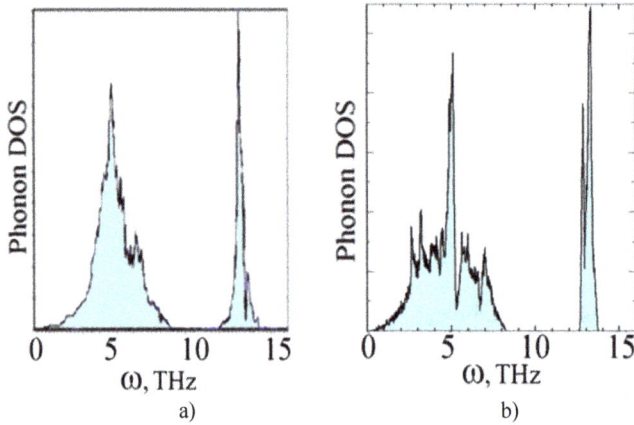

Figure 4: *Frequency distribution of atoms in the model lattice of the Pt₃Al crystal, (a) obtained directly from the model, (b) calculated according to to the classical method.*

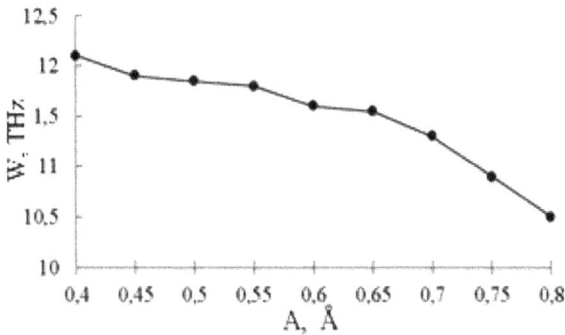

Figure 5: Dependence of the frequency of a discrete breather with a soft type of
nonlinearity on the oscillation amplitude.

The question related to the energy localized on the discrete breather is one of the
fundamental ones since this value makes it possible to judge the ability of discrete
breathers with a soft type of nonlinearity to influence the physical properties of crystals
and contribute to structural changes under intense external influences or elevated
temperatures.

The localized energy was estimated by considering the average kinetic energy of atoms
participating in concerted vibrations with the central Al atom, which has the maximum
vibration amplitude. The amount of stored energy depends on the initial deflection of the
atom. At the initial stage of oscillations, most of the energy is concentrated on one atom,
the DB nucleus. Further, the energy is redistributed to a number of neighboring atoms,
including Pt atoms.

Fig. 6 shows the dependence of the DB energy on time. This type of discrete breather can
localize energy of the order of 1 eV. Obviously, at the instant of breather destruction, this
energy is dissipated in the form of thermal vibrations to neighboring atoms.

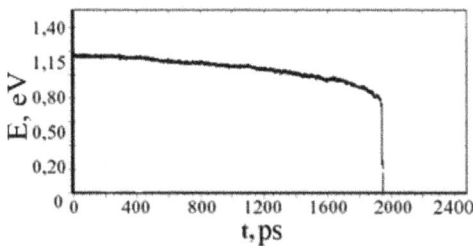

Figure 6: Dependence of the energy of a discrete breather with a soft type of nonlinearity
on time.

It is of interest to study the effect of temperature on discrete breathers in a three-dimensional crystal. For this purpose, we considered vibrations of atoms along the <100> direction with an initial deviation from the equilibrium position of 0.72 Å. The temperature varied from 0 to 150 K. The main parameter monitored was the lifetime of a discrete breather. The oscillation amplitude and frequency of an atom carrying a high-amplitude localized mode were also recorded. These parameters make it possible to estimate the dynamics of changes in the characteristics of the discrete breathers in the crystal, as well as to evaluate the possible causes of destruction of the discrete breathers.

As a result of the experiments, it was found that a slight increase in temperature leads to a significant reduction in the DB lifetime. An increase in the initial temperature from 3 to 15 K leads to a decrease in the lifetime from 420 ps to 70 ps, and a further increase in the initial temperature of the cell to 100 K and above leads to a decrease in the DB lifetime to 10 - 6 ps, which corresponds to 125 - 75 vibration periods of the atom that carries the localized mode.

The dependence of the lifetime of a discrete breather in a model Pt_3Al crystal on the initial temperature of the crystal is shown in Fig. 7. It is characteristic that at zero initial temperature, the DB lifetime is at least 2100 ps, and with a slight increase to 3 K, it is 420 ps.

As reasons leading to the destruction of the discrete breather, we consider the spread of the vibrational frequencies of the atoms entering the discrete breather, as well as the instability of the amplitude of the main carrier of the high-amplitude localized mode. Fig. 8 shows the graphs of vibrational frequencies of the Al atom in the DB core and the closest Pt atom to it in the (111) plane. It follows from the graphs that the vibrational frequency of the platinum atom is less stable than that of the Al atom. This circumstance leads to the accumulation of upward deviations from the initially established frequency and brings its value closer to the lower limit of the optical branch of the phonon spectrum of the model crystal. Ultimately, the contribution of neighboring atoms leads to an increase in the DB frequency and its entry into the phonon spectrum of the crystal, which leads to rapid energy dissipation.

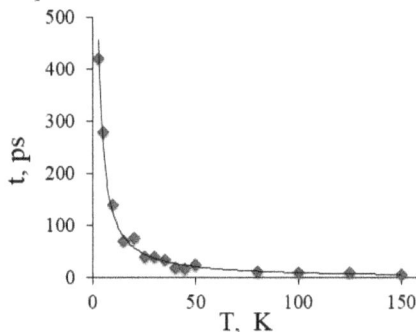

Figure 7: *Dependence of the lifetime of a discrete breather in a model Pt_3Al crystal on the initial temperature of the crystal.*

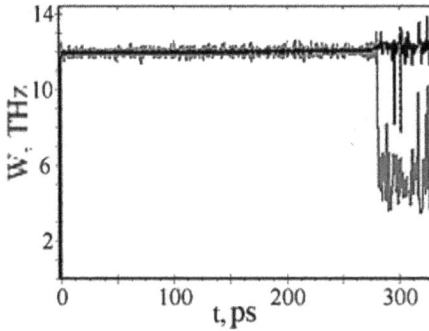

Figure 8: *Dependence of the frequency of vibrations of the Al atom (black color of the graph), which is the main DB carrier, and the nearest Pt atom (gray color of the graph) in the (111) plane in a model Pt₃Al crystal at an initial temperature of 5 K.*

Referring to the dynamics of the vibration amplitude of the Al atom in Fig. 9, we note its monotonic decrease throughout the life of the DB. An important point is that irrespective of the initial cell temperature, the final destruction of the DB occurred only after the amplitude decreased to 0.4 Å. In this case, no changes in the amplitude of vibrations of neighboring atoms, both Pt and Al, were recorded until the destruction of the DB.

Thus, we can conclude that a slight increase in the temperature of the model cell of the Pt₃Al alloy leads to a significant decrease in the DB lifetime. The decisive factor causing this is that the atoms surrounding the main atom, on which vibrations are localized, do not perform vibrations consistent with it. The higher the initial temperature of the cell, the fewer atoms are included in the composition of the DB, making its profile narrower and, accordingly, less resistant to influences from other atoms of the crystal.

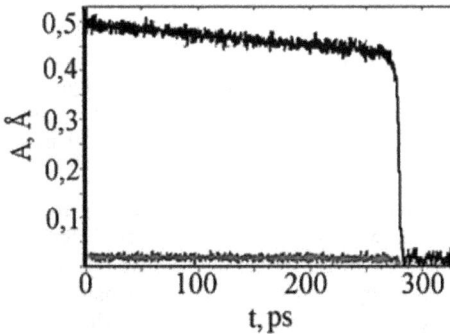

Figure 9: *Dependence of the amplitude of vibrations of the Al atom (black color of the graph), which is the main DB carrier, and the nearest Pt atom (gray color of the graph) in the (111) plane in a model Pt₃Al crystal at an initial temperature of 5 K.*

By studying discrete breathers with a soft type of nonlinearity in a Pt3Al crystal, a new type of discrete breather was discovered, which can be called an antisymmetric discrete breather (ADB), i.e., a discrete breather centered between two nodes of the crystal lattice with maximum oscillation amplitudes. ADB is an analog of the Page mode for monoatomic one-dimensional chains.

To excite ADB, two Al atoms were deflected along the <100> direction by 0.72 Å, the deflection was made towards each other or in opposite directions.

During the formation of ADB, the frequency of matched vibrations of atoms was 11.367 THz; a schematic volume profile of an antisymmetric DB is shown in Fig. 10.

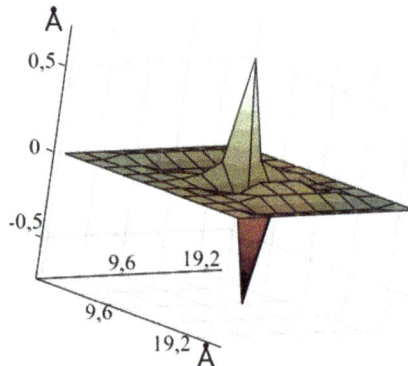

Figure 10: Schematic volumetric profile of an antisymmetric DB. The deviation of atoms of the (111) plane from the equilibrium position in angstroms is plotted along the vertical axis.

It was found that the lifetime of an antisymmetric DB depended on the initial configuration of the atoms that were removed from the equilibrium position. When setting the deviation from the equilibrium position of one atom relative to another with an accuracy of less than 5%, energy was dissipated in the crystal. The lifetime of an antisymmetric DB in our experiments did not exceed 150 ps or no more than 1750 oscillation periods.

To study the physical stability of an antisymmetric discrete breather, various initial conditions were considered when the atoms were given different initial deviations from the equilibrium position. One of the atoms in each experiment was deviated by 0.72 Å, and the second was deviated by a value of less or more than compared to the first by 0.1–6%, whereas the lifetime of the breather was measured. The results of the experiment are shown in Fig. 11.

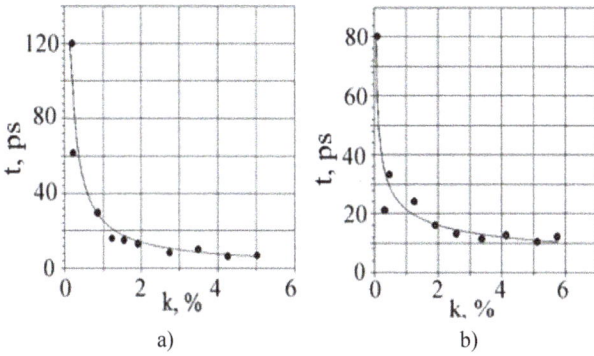

Figure 11: *Dependence of the lifetime t of an antisymmetric discrete breather on the difference in the deviation k, expressed as a percentage, of one atom relative to another. Figure (a) the second atom had a downward deviation; (b) the second atom had a larger deviation.*

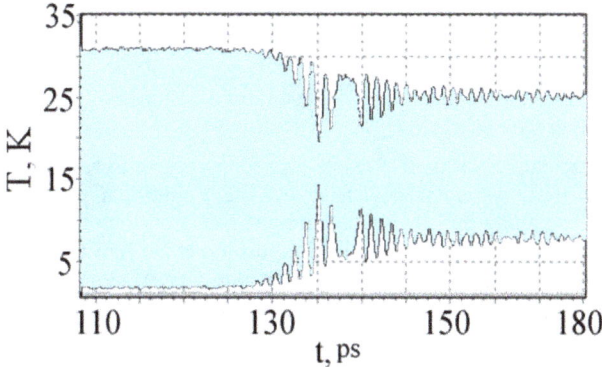

Figure 12: *Temperature curves of the Al and Pt sublattices during the tuning of ADB vibrations in a DB with a soft type of nonlinearity.*

Note that, in our case, the destruction of the antisymmetric breather did not lead to the complete dissipation of energy over the Pt_3Al crystal. As a result of the destruction, a DB with one atom in the nucleus was formed, the frequency of atomic vibrations was 10.526 THz, and the amplitude of vibrations was 0.745 Å. In this case, the lifetime of steady-state lattice oscillations exceeded 1500 ps or more than 15500 oscillation periods.

The process of tuning the oscillations of an antisymmetric discrete breather in the DB takes 10 – 15 ps (Fig. 12). Part of the energy is dissipated into the Al sublattice without being transferred to the Pt sublattice. This can be observed in the temperature curves in Fig. 12, where this process can be seen from 128 ps to 146 ps.

Thus, we can speak of at least two stable types of discrete breathers with a soft type of nonlinearity in the Pt_3Al crystal.

2.2 Discrete breathers with a hard type of nonlinearity in a 3D model of Pt3Al

In addition to energy localization, energy transport through the crystal is of considerable interest. Energy transfer in the crystals under consideration using DBs with a soft type of nonlinearity is impossible due to its localization predominantly on one atom, however, DBs, localized on a group of atoms, may have the possibility of energy transfer.

The search for discrete breathers with a hard type of nonlinearity was carried out along close-packed directions of the Pt_3Al crystal. Such discrete breathers can be excited in a crystal by setting a group of n atoms with certain initial displacements

$$r(t) = \{r(t), r_2(t), \dots, r_n(t)\}, \tag{3.9}$$

or initial velocities.

After a series of experiments, it was possible to excite a discrete breather localized mainly on Al atoms. In contrast to DBs with a soft type of nonlinearity, the vibrations of DBs with a hard type involve several atoms of the light sublattice. DBs with a hard type of localization can move through the crystal for considerable distances, practically without dissipating their energy.

The following technique was used to excite a stationary discrete breather with a hard type of nonlinearity. Two Al atoms deviated from the equilibrium position by 0.9 Å in opposite directions along the <110> direction, as shown in Fig. 12a (numbers 1 and 2 denote the atoms taken out of the equilibrium position at the zero moments). As a result of the transient process, the energy is redistributed to four Al atoms (Fig. 13b). The DB excitation is possible along close-packed directions: <110>, $< 0\bar{1}1 >$, $< \bar{1}01 >$, and $< \bar{1}10 >$.

After the transient process, DBs can exist for a long time and move along the indicated crystallographic directions. Fig. 14 shows the stroboscopic picture of such a DB, as well as the 3D energy profile of the breather. This allows us to evaluate its localization, both spatial and energy.

The most important characteristics of discrete breathers include their lifetime in a crystal, the amount of energy they can localize, the degree of spatial localization, and the frequency dependence on the oscillation amplitude. Let's consider these characteristics in more detail.

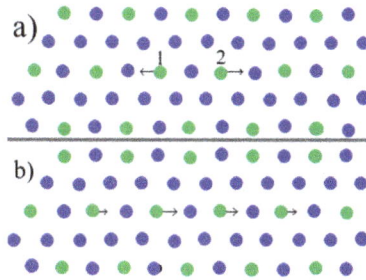

Figure 13: *Plane (111) of a Pt₃Al crystal: (a) Initial conditions for excitation of a discrete breather with a hard type of nonlinearity along the $< \bar{1}10 >$ direction; (b) Formed DB that appeared after the transient process after 2 ps.*

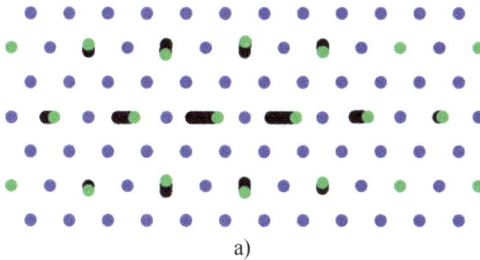

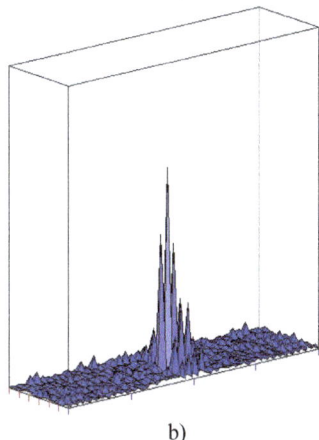

Figure 14: *Discrete breather with a hard type of nonlinearity. (a) The stroboscopic pattern of a discrete breather with a hard type of nonlinearity, (b) three-dimensional DB profile in the plane.*

First of all, let us consider the spatial localization of DBs since this parameter plays a significant role in studying the stability of DBs to various influences. For example, with an increase in temperature, the DB lifetime significantly decreases.

The DB is mainly localized on four Al atoms (Fig. 15) and extended along one of the close-packed rows of the crystal. Note that the atoms oscillate along the close-packed row in which the discrete breather is excited. Possessing mobility, the DB is less stable compared to the DB with a soft type of nonlinearity due to its smaller localization.

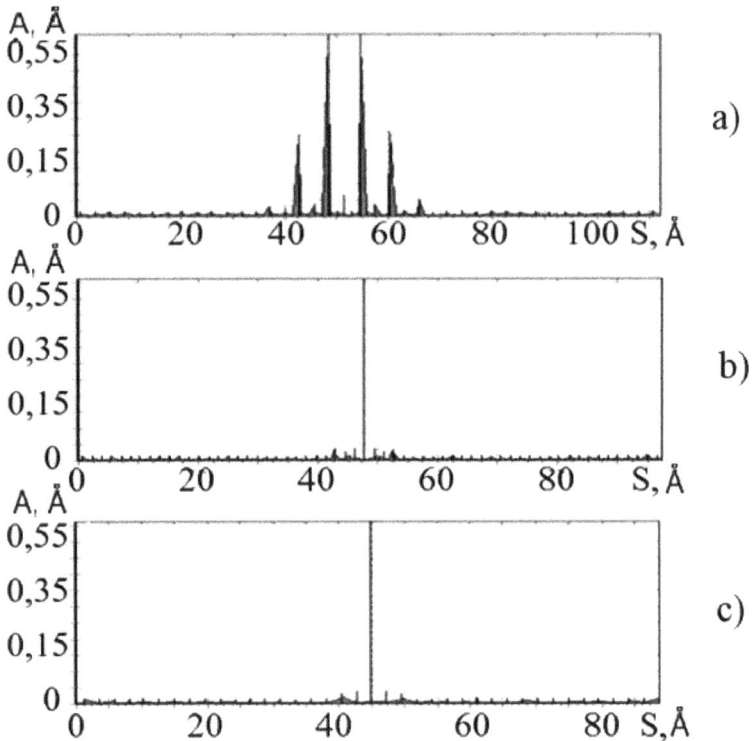

Figure 15: *Discrete breather profile projection: (a) onto the (111) plane; (b) onto the (1\overline{1}0) plane; (c) onto the (11\overline{3}) plane. The amplitude of atomic vibrations in Å is plotted along the ordinate axis, and the distance in the corresponding direction is plotted along the abscissa axis in Å.*

As in the case of soft nonlinearity, the dependence of the stored energy on time was obtained (Fig. 16). Such breathers are already capable of localizing energy of the order of 2–3 eV, which, undoubtedly, can have a local effect on defects in the case of destruction of discrete breathers near them.

The dependence of the frequency on the oscillation amplitude is compared for both types of discrete breathers. This graph clearly demonstrates the difference between DBs with hard and soft types of nonlinearity. In this case, the frequency increases with increasing vibration amplitude and lies above the optical branch of the phonon spectrum of Pt_3Al crystal.

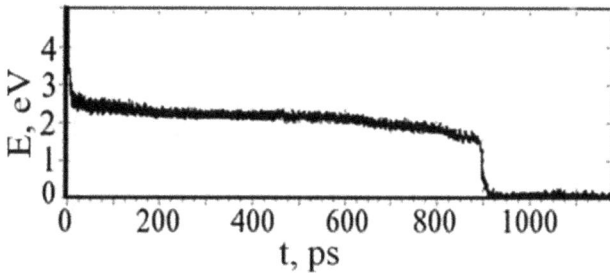

Figure 16: *Dependence of the energy of a discrete breather with a hard type of nonlinearity on time.*

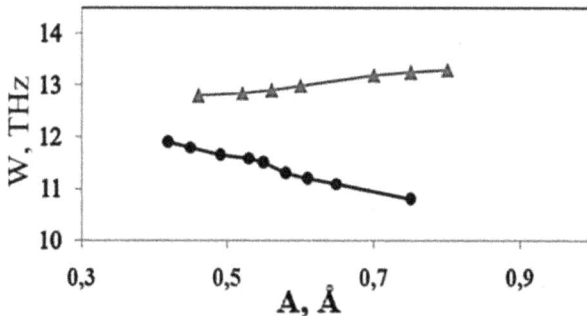

Figure 17: *Dependence of the frequencies of the DB with a soft type of nonlinearity (round marker) and DB with a hard type of nonlinearity (triangular marker) on the oscillation amplitude.*

Consider a mobile DB. To excite it, two Al atoms deviated from the equilibrium position by 0.9 and 0.85 Å in opposite directions along the close-packed direction, as shown in Fig. 13a. As a result of the transient process, the energy is redistributed to four Al atoms (Fig. 13b). The resulting DB has an initial momentum towards the atom with a smaller initial deviation.

To consider a moving DB, we used an enlarged cell along the direction of the breather movement with a size of 225.71 x 29.32 x 20.73 Å.

Let us analyze the dynamics of a moving breather with a hard type of nonlinearity in a Pt_3Al crystal. An important characteristic of a discrete breather is its lifetime and the maximum distance it can travel along the crystal in the direction $< \overline{1}10 >$. Fig. 18 shows the dependence of the speed of movement of the DB on the distance traveled by it, and Fig. 19 shows the dependence of the oscillation amplitude on the distance traveled.

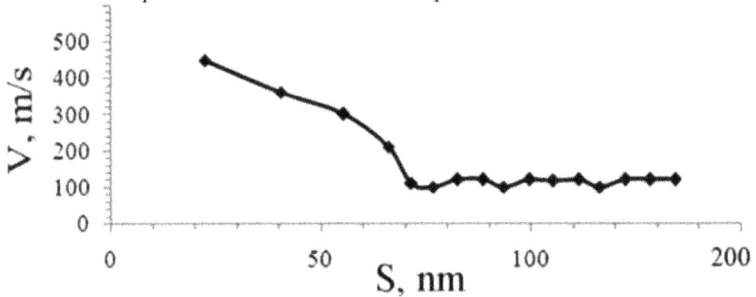

Figure 18: *Dependence of the DB velocity on the distance traveled by it in the Pt_3Al crystal along the direction $< \overline{1}10 >$.*

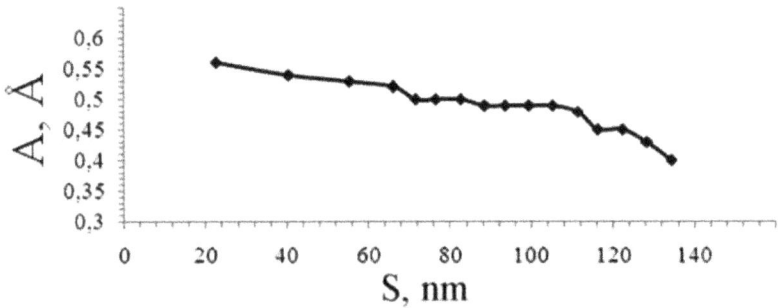

Figure 19: *Dependence of the amplitude of vibrations of DB atoms on the distance traveled by them in a Pt_3Al crystal along the direction $< \overline{1}10 >$.*

Due to the periodic boundary conditions, the DB can pass through the computational cell an unlimited number of times. Note that the initial speed of the DB movement through the crystal reaches 450 m/s, however, later it decreases and approaches a value of 110 – 120 m/s, while the breather slowly dissipates energy into the Al sublattice. The change in the amplitude of vibrations occurs smoothly during the entire time of its existence, but a decrease in the amplitude of vibrations of atoms to 0.4 Å leads to its destruction. This can be explained by the fact that as the amplitude of the discrete breather decreases, its frequency also decreases and enters the optical branch of the phonon spectrum, which leads to the excitation of phonons and the rapid dissipation of energy into the light sublattice of the crystal.

Thus, it has been shown that a breather with a hard type of nonlinearity is capable of moving over a considerable distance in a Pt_3Al crystal. In this case, its destruction leads to local heating of the Al sublattice, followed by energy dissipation throughout the crystal.

Next, we turn to the question of the interaction of discrete breathers in a three-dimensional Pt_3Al crystal. The theory of solitons, developed for integrable nonlinear equations, describes their elastic interaction. However, this work considers quasi-breathers, and absolutely elastic interactions between discrete breathers were not observed.

Let us consider the interaction of two mirror-symmetric breathers with a rigid type of nonlinearity moving toward each other. During the first collision of DBs, they are elastically repulsed from each other (see Fig. 20). Upon collision, part of their energy is scattered into the Al sublattice. This is clearly seen when comparing Figs. 20a and 20d, which shows that the DB amplitudes before the collision were 0.6 Å, and after the collision, 0.45 Å. Repeated collision of the DB leads to the destruction of one of them. In this process, not only a decrease in the amplitude of the breather oscillations plays a role, but also the heating of the Al sublattice during their movement through the crystal.

Considering the collision of the DBs with a hard type of nonlinearity and with a soft one (see Fig. 21), we pay attention to the high stability of a stationary breather polarized along the <100> direction. As a result of the collision, the DB with a hard type of nonlinearity slowed down its speed to 120 m/s, and its amplitude decreased to 0.55 Å. Repeated collisions did not lead to the same significant energy losses, but its lifetime did not exceed 350 ps. A DB with a soft type of nonlinearity practically did not lose its energy as a result of collisions and could exist for 2000–2500 ps.

In the collision of two or more discrete breathers with a hard type of nonlinearity moving along different crystallographic directions, their mutual configuration during the collision is important. If the breathers reach the meeting point at the same time, then, as in the case of moving along one direction, they mutually repulse with the loss of part of the energy, but a repeated collision, in this case, is impossible due to their movement along different atomic rows. If one of the breathers reaches the DB meeting point earlier, then the second "hit" it, which leads to the destruction of the first DB, and the second changes its direction of movement to the opposite.

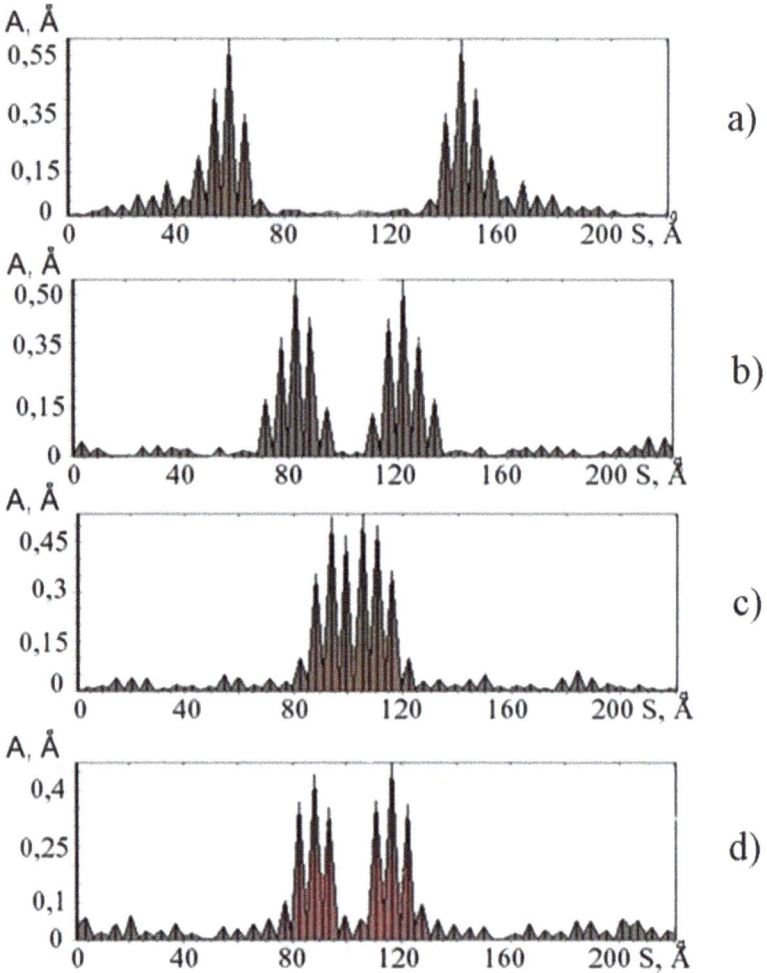

Figure 20: *Collision of two mirror-symmetric discrete breathers with a hard type of nonlinearity moving towards each other along the direction* $< \overline{1}10 >$ *of the Pt₃Al crystal.*

The distance along the direction $< \overline{1}10 >$ *is plotted along the abscissa axis and the amplitude of atomic vibrations in angstroms along the ordinate axis. (a) DB formation 4 ps after the start of the experiment. (b) Rapprochement of DB after 8 ps. (c) Collision of two DBs after 10 ps. (d) The divergence of two DBs in different directions after the collision after 16 ps from the beginning of the experiment.*

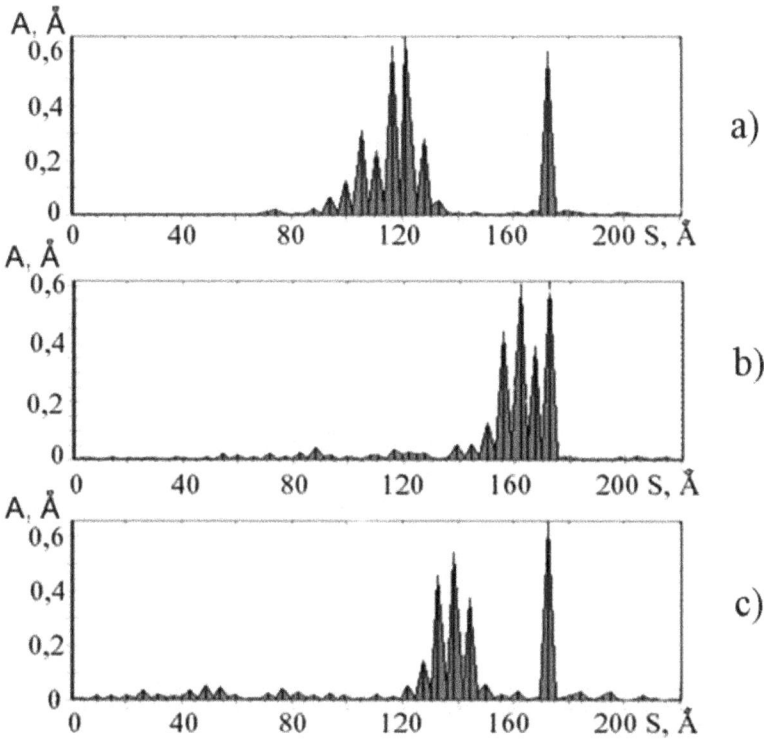

Figure 21: *DB interaction in a Pt₃Al crystal. (a) The initial moment of DB formation is 4 ps after the beginning of the experiment. (b) The moment of collision of two DBs after 10 ps of the experiment. (c) Reflection of the DB after 16 ps from the beginning of the experiment.*

2.3 Interaction of a discrete breather with a hard type of nonlinearity with a vacancy

Next, we consider the interaction of a mobile discrete breather with a rigid type of nonlinearity with a vacancy, depending on the initial conditions for the formation of the breather. This discrete breather was excited in the same way as in the previous section, at a distance of 32 Å from the vacancy, to avoid its influence at the initial stages of motion. We have obtained the values of the energy of a moving discrete breather during the period of collision with a point defect. Also considered is the dependence of the time of motion of the DB over the crystal, the frequency and amplitude of oscillations of atoms located near a point defect, on the deviations of one of the Al atoms (0.5 ÷ 1 Å) at the moment of launching the DB.

Fig. 22 shows the dependence of t (time of motion of the DB before the collision with a defect in ps) on Δ (deviations of the abscissa of the right Al atom in Å at a fixed deviation of the left Al atom of 1 Å from the equilibrium position). This graph makes it possible to estimate the influence of the initial conditions of the formation of a breather on the speed of its movement and, accordingly, to characterize the interaction of the breather with a vacancy on the average speed of its movement through the crystal.

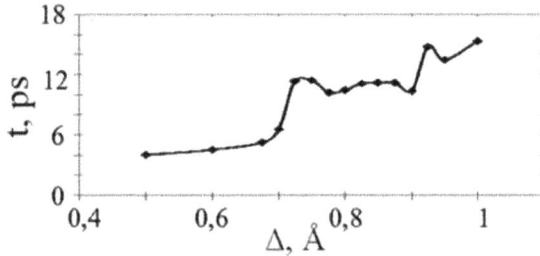

Figure 22: Dependence t(∆).

Fig. 23 shows the dependence of the DB energy E on Δ and Fig. 24 shows the dependence of the DB velocity v on Δ.

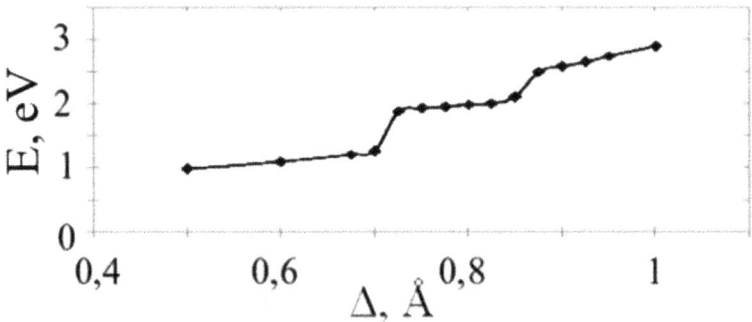

Figure 23: Dependence ε(∆).

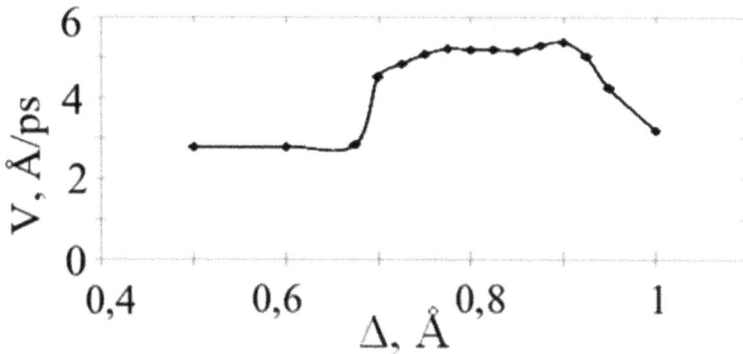

Figure 24: Dependence V(Δ).

Based on the given dependencies and the results of computer experiments, several regularities can be identified, which are as follows. Firstly, when changing the deviation Δ of the abscissa of the right Al atom from the equilibrium position by $0.825 \div 1$ Å, the moving DB reaches the defect, collides elastically with it, interacts with the defect on the order of 0.63 ps, and moves in the opposite direction, while interaction gives part of its energy, about 0.113 eV, to the Al sublattice. The energy of the moving DB during the period of collision with a defect varies from 2 to 2.9 eV. When the moving DB interacts with a point defect, it gives up about 5% of its energy to the defect and then continues its movement in the opposite direction. The frequency of atomic vibrations near a point defect varies from 12.65 to 12.85 THz, and their amplitude varies from 0.025 to 0.08 Å. With this interaction of the moving DB with a defect, its motion velocity varied from 3.19 to 5.20 Å/ps. Secondly, when changing the deviation Δ of the abscissa of the right Al atom from the equilibrium position by $0.725 \div 0.8$ Å, the moving DB reaches the defect, collides with it, interacts with the defect of the order of 1.26 ps, and is destroyed. The frequency of atomic vibrations near a point defect varies from 12.58 to 12.67 THz, and their amplitude is from 0.03 to 0.045 Å. With this interaction of the moving DB with a defect, its motion velocity varied from 4.85 to 5.20 Å/ps. Thirdly, when changing the deviation Δ of the abscissa of the right Al atom from the equilibrium position by $0.5 \div 0.7$ Å, the moving DB travels a distance that varies from 21.8 to 39.83 Å and is destroyed, it does not reach the defect and therefore does not interact with it. The energy of the moving DB during this interaction varies from 1.25 to 0.98 eV, and its movement velocity varies from 2.8 to 4.51 Å/ps. The frequency of atomic vibrations near a point defect varies from 3.5 to 5.5 THz, and their amplitude from 0.005 to 0.028 Å.

Table 2 lists the average values of the rate of motion of a discrete breather over a crystal, the mean energy of a moving breather in a collision with a defect, and the mean amplitude and frequency of atomic vibrations near a point defect. Analyzing the tabular values, we can say that at an average speed of motion of the moving DB over a crystal of about 4.79 Å/ps, an elastic interaction of the moving DB with a point defect is observed.

The moving DB reaches the defect, interacts with it, and moves back. In this interaction, the average amplitude and frequency of atomic vibrations near a point defect are 0.058 Å and 12.76 THz, respectively.

Table 2: *Average values of numerical characteristics of a moving discrete breather.*

Δ – deviation of atom Al in Å	0.825÷1.0	0.725÷0,8	0.5÷0.7
U_{cp} – average velocity of DB movement along the crystal, in Å/ps	4.79	5.09	3.38
ε_{cp} – average DB energy upon collision with a defect in eV	2.5	1.93	1.13
A_{cp} – average amplitude of atomic vibrations near a defect in Å	0.058	0.044	0.015
υ_{cp} – average frequency of vibrations of atoms near a defect in THz	12.76	12.63	4.58

If the speed of the moving DB reached 5.09 Å/ps, then the DB, having reached the defect, is destroyed upon interaction with it. In this interaction, the average amplitude and frequency of atomic vibrations near a point defect are 0.044 Å and 12.63 THz, respectively. A slight decrease in the amplitude and frequency of oscillations of nearby atoms with a point defect is obviously because the average energy of a moving DB when interacting with a point defect is 2.5 eV, and for the case when the DB is destroyed, it is 1.93 eV. When the speed of the moving DB decreases to 3.38 Å/ps, the DB does not reach the defect and does not interact with it, but is destroyed and dissipates energy over the crystal in the form of thermal vibrations.

2.4 The influence of deformation on the characteristics of DBs with soft and hard types of nonlinearity

Structural and functional materials are often subjected to intense external influences, which manifest themselves in material deformations. Thus, the problem arises of studying the effect of elastic deformation of all-round tension/compression on the characteristics of DBs in the considered model of the Pt_3Al alloy.

The three-dimensional calculation cell of Pt_3Al contained 7200 atoms and the Morse potential was used. The initial conditions for the excitation of DBs with a soft type of nonlinearity were the same as described above. We carried out elastic deformation of all-round tension/compression of the Pt_3Al crystal cell, the value of this deformation varied from 1 to 10%.

Experiments have shown that crystal deformation leads to significant changes in the characteristics of discrete breathers. Figs. 25 and 26 show the dependences of the oscillation frequency of the DB, as well as its lifetime, on the magnitude of the elastic deformation of all-round tension/compression.

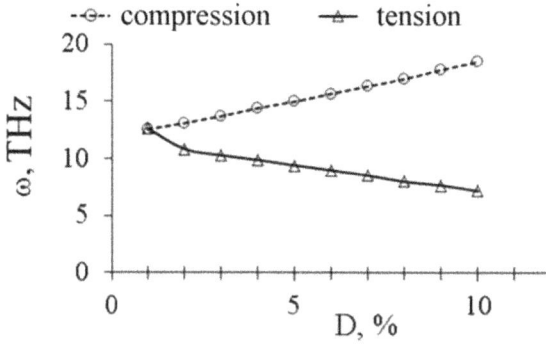

Figure 25: Dependence of the oscillation frequency ω (in THz) of a discrete breather on the value of the elastic deformation of the all-round tension-compression D (in %) of a Pt₃Al crystal cell.

An increase in the compressive elastic strain leads to an increase in the frequency of the DB oscillations, and an increase in the tensile elastic strain leads to a decrease in the DB oscillation frequency.

Note that the DB frequency as a function of deformation can be approximated by a linear dependence with good accuracy. For elastic tensile strain, this dependence has the form:

$$\omega = -0{,}5295D + 12{,}27, \tag{18}$$

and for elastic compressive deformation:

$$\omega = 0{,}6627D + 11{,}768. \tag{19}$$

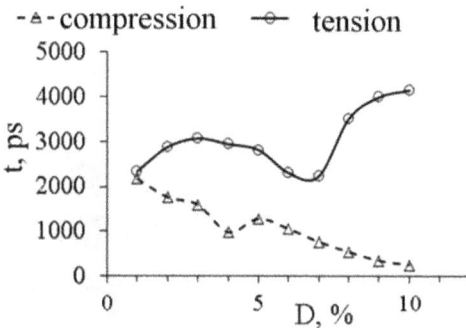

Figure 26: Dependence of the lifetime t (in ps) of a discrete breather on the value of the elastic deformation of all-round tension-compression D (in %) of a Pt₃Al crystal cell.

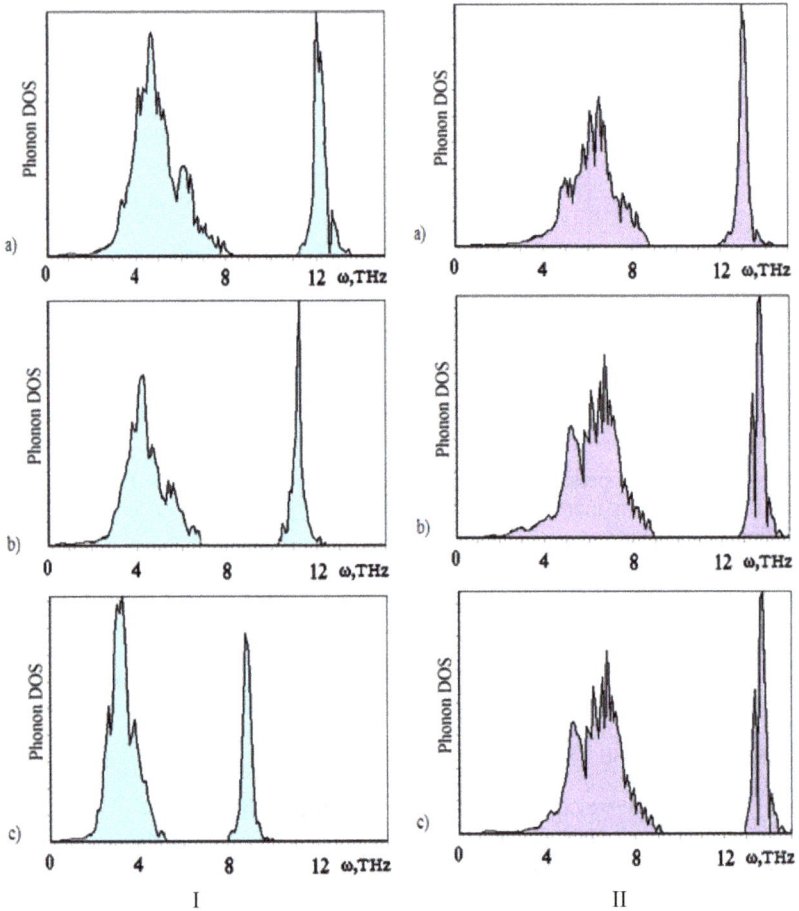

Figure 27: *Densities of phonon states of a Pt₃Al crystal cell, I) under elastic tensile strain: (a) 1% extension, (b) 3% extension, (c) 8% extension; II) under elastic deformation of compression: (a) compression 1%, (b) compression 2%, (c) compression 3%.*

The elastic deformation of all-round tension/compression of a Pt_3Al crystal cell significantly affects the DB lifetime. Noting the general nature of the dependence of the DB lifetime on the elastic strain (see Fig. 26), we can say that an increase in the compressive elastic strain decreases the DB lifetime, and an increase in the tensile elastic strain increases the DB lifetime. For example, for an elastic tensile strain of 10%, the DB lifetime is about 4150 ps, and for the same value of elastic compressive strain, about 258 ps.

For the existence of discrete breathers, the distribution of phonon modes is of primary importance, i.e., the phonon spectrum of the crystal.

To study the effect of deformation on the phonon spectrum of the crystal, the cell was heated to 5 K, then the vibration frequencies of the cell were recorded, and the corresponding distribution was plotted. This approach is preferable to the theoretical calculation of the phonon spectrum because is provided by data directly from the model at finite low temperatures, which cannot always be taken into account in a deformed crystal.

The presence of a local minimum on the curve for the dependence of the effect of tensile strain on the lifetime may indicate a characteristic change in the phonon spectrum of the crystal at a given strain value. Fig. 27 shows the phonon density of states of a cell of a Pt_3Al crystal at various values of tensile and compressive elastic strains.

Under compressive deformation, the optical branch of the phonon spectrum shifts towards higher frequencies, while under tensile deformation, it shifts towards lower frequencies.

Figs. 28 and 29 show the dependences of its energy, as well as the amplitude of oscillations of the DB, on the magnitude of the elastic deformation of the all-around tension-compression.

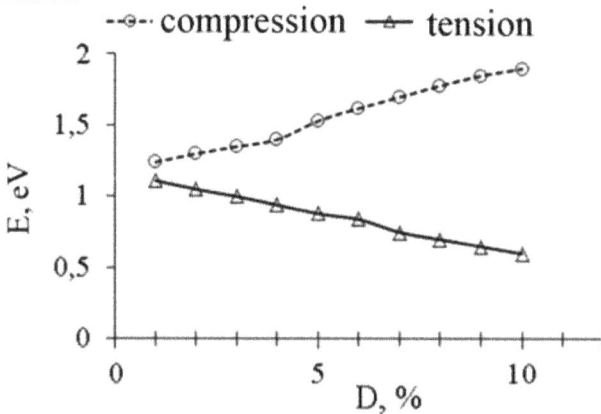

Figure 28: *Dependence of the energy E (in eV) of a discrete breather on the value of the elastic deformation of the all-round tension-compression D (in %) of a Pt_3Al crystal cell.*

It can be seen in Fig. 28 that with an increase in the elastic tensile strain, the DB energy decreases, and with an increase in the elastic compressive strain, it increases. By analogy with the analytical asymptotic estimate of the frequency, the DB energy as a function of deformation can be approximated with good accuracy by a linear dependence. For elastic tensile strain, this dependence has the form:

$$E = -0{,}0576D + 1{,}1687, \tag{20}$$

and for elastic compressive deformation:

$$E = 0{,}0784D + 1{,}136. \tag{21}$$

It can be seen in Fig. 29 that with an increase in the elastic tensile strain, the amplitude of oscillations of the DB increases, and with an increase in the elastic compressive strain, it decreases. The linear approximation of the amplitude as a function of elastic tensile strain has the form:

$$A = 0{,}0111D + 0{,}534, \tag{22}$$

and for elastic compressive deformation:

$$A = -0{,}0141D + 0{,}5433 \tag{23}$$

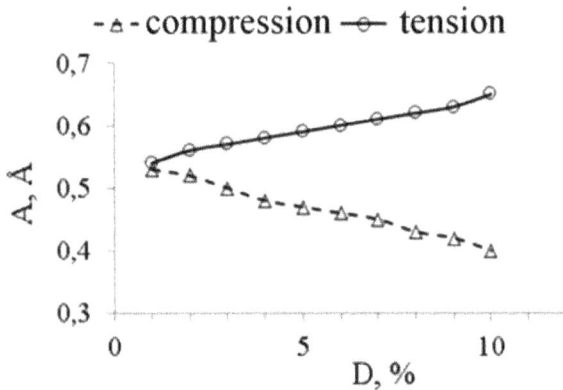

Figure 29: *Dependence of the oscillation amplitude A (in Å) of a discrete breather on the value of the elastic deformation of all-round tension-compression D (in %) of a Pt₃Al crystal cell.*

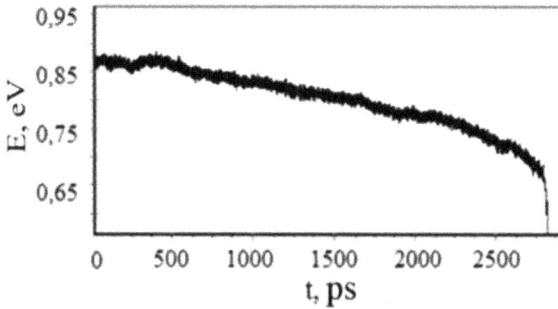

Figure 30: *Dependence of the energy E (in eV) of a discrete breather on the calculation time t (in ps) for an elastic tensile strain of 5%.*

As an example, Fig. 30 shows the dependence of the energy E (in eV) of DB on the calculation time t (in ps) for an elastic tensile strain of 5%.

The obtained dependencies of the energy of a discrete breather on the magnitude of the strain also show that the energy of the discrete breather increases under the uniform compression strain, and decreases under the tensile strain. The data obtained indicate that the ratio of the DB energy at which it is destroyed to the initial energy is preserved and amounts to a value of the order of 0.77 - 0.8. This means that the destruction of the DB occurs after the dissipation of 20 - 23% of the energy during the time of its existence, and the rest is dissipated in the process of destruction of the DB.

Potential prospects for the use of discrete breathers with a hard type of nonlinearity in various systems, including nanofibers, require a more detailed study of the influence of various factors on the possibility of excitation of moving breathers, as well as on their characteristics.

In this section, we study the influence of elastic deformation on the possibility of excitation of a movable discrete breather in an A_3B crystal, using the Pt_3Al crystal as an example.

The model under consideration was a bulk crystal of A_3B stoichiometry, 225.71 x 29.32 x 20.73 Å in size, containing 8640 particles, the boundary conditions were set periodic along the <110> direction and free in other directions, the atoms interacted through the Morse pair potential.

The DB excitation technique corresponds to that described above and consisted in giving two atoms of the light sublattice initial deviations from the equilibrium position in opposite directions. The deviation value was set in such a way that after relaxation the resulting DB in an undeformed crystal had a maximum velocity.

The nanofiber deformation was set following the Poisson rule. The main attention was paid to the direction along which the DB moved. This direction corresponded to the crystallographic direction $< 1\overline{1}0 >$.

Then the other two perpendicular directions were subjected to deformation in line with the following formula:

$$\varepsilon_{<111>} = \varepsilon_{<1\overline{1}2>} = -\mu\varepsilon_{<1\overline{1}0>}, \tag{24}$$

where $\varepsilon_{<111>}$, $\varepsilon_{<1\overline{1}2>}$, $\varepsilon_{<1\overline{1}0>}$ are the corresponding relative strains along crystallographic directions, μ is Poisson's ratio.

Crystal deformation leads to significant changes in its properties. For the existence of discrete breathers, the distribution of phonon modes is of primary importance, i.e., the phonon spectrum of the crystal. To study the effect of deformation on the PS of the crystal, the cell was heated to 5 K, then the vibration frequencies of the cell were recorded, and the corresponding distribution was plotted, as in the previous section. This approach is preferable to the theoretical calculation of the PS because is provided by data directly from the model at finite low temperatures, which cannot always be taken into account in a deformed crystal.

Fig. 31. shows the PS for various cell deformations. The obtained results of the influence of elastic deformation somewhat differ from theoretical calculations for other crystals, however, it should be noted that the trend is similar. Under compressive strain, the optical branch of the PS shifts towards higher frequencies, while under tensile strain, it shifts towards lower frequencies.

Investigating the effect of such deformations on the conditions of excitation of DBs, it was found that even a slight deformation significantly affects the characteristics of the obtained breather.

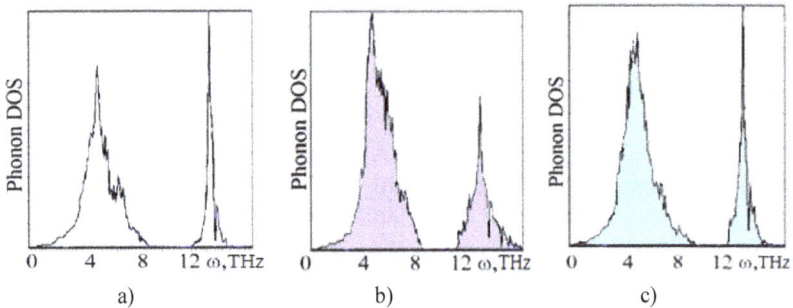

Figure 31: *Phonon spectrum of a Pt₃Al crystal, (a) no deformation, (b) compression strain 3%, (c) tensile strain 3%.*

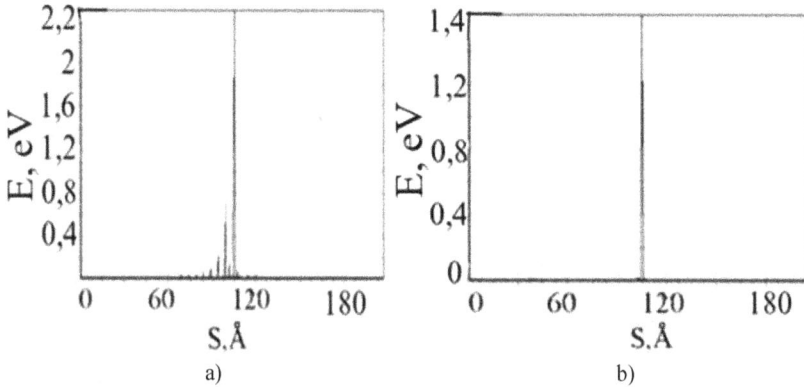

Figure 32: *Energy distribution of atoms along the nanofiber, (a) the initial stage of the experiment 5 ps, (b) the formed DB with a soft type of nonlinearity after 15 ps.*

With a compression strain along $< \overline{1}10 > 0.5\%$, a stable moving DB is formed, its velocity $v = 2.04$ A/ps, and its frequency $\omega = 13.69$ THz. At a compression strain of 1%, a stable moving DB is formed, its velocity $v = 1.17$ A/ps, and its frequency $\omega = 13.33$ THz. Further deformation of the cell did not lead to the formation of conditions for the excitation of a moving discrete breather. But at a compression strain of 1.5% to 3%, an unstable nonlinear vibrational mode is formed, which exists on the order of 6 ps, then a DB is formed with a soft type of nonlinearity, with a frequency falling into the forbidden gap in the phonon spectrum, and its lifetime is more than 30 ps.

Fig. 32 shows a breather with a hard type of nonlinearity turning into a DB with a soft one. At compressive strains of more than 3%, no mobile DB is formed, and the formation of stationary discrete breathers was also not observed.

Tensile strain along the nanofiber led to conditions where no DB was formed even at strains less than 0.5%. At a tensile strain of 2.5%, an unstable nonlinear vibrational mode is formed, which exists for about 5 ps, then a DB is formed, with a frequency $\omega = 10.9890$ THz, which falls into the band gap of the PS. The lifetime of such a DB was more than 40 ps. Note that the direction of the DB polarization corresponds to the direction $< \overline{1}10 >$, and the DB is polarized along the direction $<100>$, thus, during the formation of a DB with a soft type of nonlinearity, a change in the polarization of oscillations of the nonlinear localized mode occurs.

2.5 Excitation of discrete breathers under intense external actions in A₃B crystals

Intense external influences on crystals lead to significant deviations of atoms from lattice positions, activating various processes that cannot be studied within the framework of linearized equations of motion. Such processes can lead to the formation of vacancies,

Frenkel pairs, and other defects in crystals. Anharmonisms of interatomic bonds are responsible for various nonlinear phenomena, ranging from thermal expansion of crystals, excitation of localized vibrational modes, and ending with structural rearrangements in a crystal.

In this section, we consider possible mechanisms of excitation of discrete breathers with a soft type of nonlinearity in A_3B crystals with an $L1_2$ superstructure under intense external influences.

Following modern ideas about the dynamics of a crystal lattice and defect formation under intense external influences, two variants of excitation of the DBs under external influences on a Pt_3Al crystal can be considered, - a direct collision of a fast particle with a lattice atom or a more complex process associated with the excitation of the electronic subsystem of the crystal. The first case is realized for particles that carry a noticeable momentum, that is, for electrons, ions, and neutrons. Electromagnetic radiation quanta, even as energetic as γ-quanta, do not directly displace atoms from lattice sites. However, by transferring their energy to electrons, they can initiate significant deviations of atoms from the equilibrium position, which in some cases can lead to the formation of defects.

We consider the mechanism of excitation of discrete breathers during the interaction of atoms with particles carrying a noticeable momentum, for example, neutrons. With this mechanism, the displacement of an atom occurs so quickly that its environment does not have time to rearrange itself, and due to the anharmonicity of interatomic bonds, it is possible to form conditions for the excitation of discrete breathers with a mild type of nonlinearity.

To simulate an external intense action on the crystal, randomly selected atoms in Pt_3Al were subjected to a momentum along the possible direction of excitation of a DB with a soft type of nonlinearity, thereby simulating the interaction of atoms, for example, with neutrons or high-energy electrons. The number of atoms receiving momentum depended on the energy transferred to the atoms. The pulse energy range varied from 0.2 to 5 eV per interaction, respectively, from 1.5% to 0.025% of the total number of atoms in the calculated cell received it. The pulse periodicity varied from 0.5 ps to 10 ps. Thus, a heating process was provided that did not lead to the destruction of the model. Such heating can be observed, for example, during the electropulse processing of metals.

We have found that the minimum energy E_{db} of the DB with a soft type of nonlinearity for a Pt_3Al crystal is about 0.8 eV; however, more energy is required to excite a DB, because some of the energy is dissipated in the lattice. Thus, the Al atom must accumulate the required amount of energy due to one or more received impulses.

The energy E transferred to a lattice atom can be calculated from the laws of conservation of momentum and energy, assuming that the impact is fully elastic. The value of E will be maximum at a central collision, and for non-relativistic particles:

$$E = \frac{4M_1M_2}{(M_1+M_2)^2}E_0, \tag{25}$$

where M_1 and M_2 are the masses of the incident particle and the lattice atom, respectively, E_0 is the energy of the particle, and the lattice atom is assumed to be at rest.

Estimation of the energy transferred to atoms is important from the point of view of studying discrete breathers in full-scale experiments. Knowing the energy of the discrete breather and assuming that $E_{db} = E$ from Eq. (25), it is possible to select the energy range of the incident particles and reduce the search time for suitable parameters. Or, by sorting through the energies of incident particles and tracking the parameters of the sample under study, one can reveal the contribution of discrete breathers to certain properties of the crystal.

Fig. 33 shows the generation of discrete breathers in a bulk Pt_3Al crystal at $E = 5$ eV. This energy can be obtained by an aluminum atom in a collision with a neutron with an energy E_0 of the order of 35 eV. Note that the pulse was transmitted every 5 ps to 0.05% of the atoms of the computational cell.

In the case of a decrease in the energy transferred to atoms and an increase in the percentage of atoms that received a pulse, the effect of energy accumulation in the light Al sublattice is possible, leading to the excitation of discrete breathers. As an example, Fig. 34 shows the temperature curves of the sublattices of the Pt_3Al alloy.

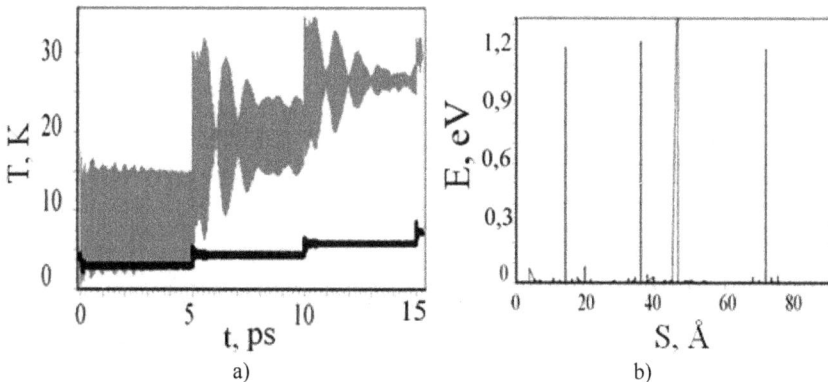

Figure 33: (a) Temperature curves of sublattices Pt (black color of the graph) and Al (gray color of the graph) during interaction with particles with an energy of 5 eV; (b) Energy distribution along the crystal at 9 ps.

Thus, it is possible to reveal two ways of excitation of discrete breathers with a soft type of nonlinearity in a Pt_3Al crystal under the action of high-energy particles on it. In the first case, the discrete breather is excited due to a single collision, and in the second, due to the accumulation of energy on the Al atom due to a series of collisions.

By varying the parameters of the impact on the model crystal, we obtained the dependences of the time of the first fixation of the DB on the amount of energy transferred to the atoms, as well as on the percentage of atoms receiving an impulse, at a fixed interaction energy.

As can be seen from the graph in Fig. 33a, the mechanism of energy accumulation is realized at energies from 0.4 to 1.4 EV; For energies from 1.4 to 2.2 EV and further up to 5 eV, DBs can arise in a single collision or 2-3 collisions. At energies less than 0.4 eV, discrete breathers were not detected. This is because to excite a DB, an aluminum atom needs to receive at least 6-8 successive collisions at such energies, leading to an increase in the amplitude of its oscillations. An increase in the cell temperature with each subsequent pulse also plays a negative role. It should be noted that the lifetime of DBs generated in this way is not long: DBs exist from 7 to 15 oscillation periods under such conditions, which corresponds to 0.56–1.12 ps.

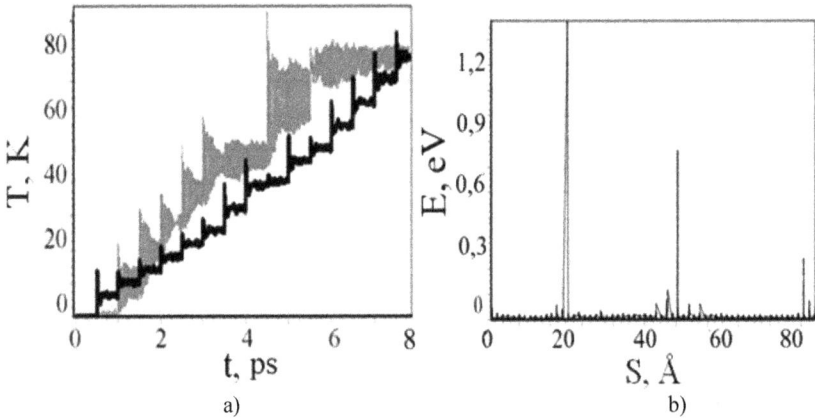

Figure 34: (a) Temperature curves of Pt (black color of the graph) and Al (gray color of the graph) sublattices during interaction with particles with an energy of 1 eV on the crystal lattice of the Pt_3Al alloy; (b) Energy distribution along the crystal at 5 ps.

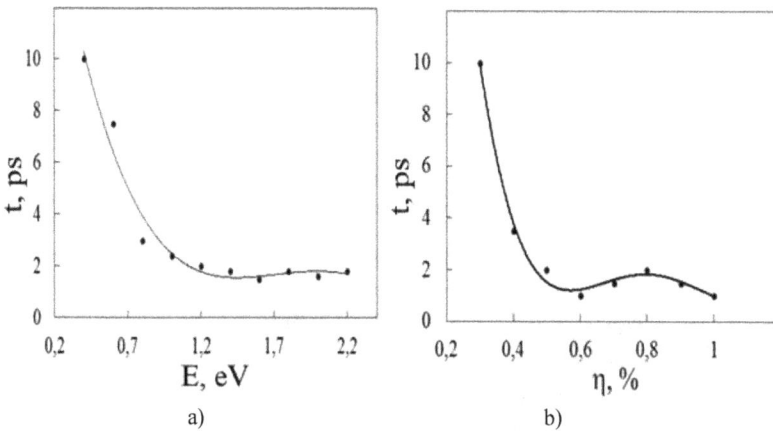

Figure 35: *(a) Dependence of the time of the first fixation of the DB in the computational cell on the energy transferred to the atom in one act of interaction with the incident particle. The pulse was transmitted every 0.5 ps, 0.5% to the atoms in the cell; (b) Dependence of the time of the first fixation of DBs in the computational cell on the percentage of atoms interacting with incident particles. The pulse was transmitted every 0.5 ps, the interaction energy was 1 eV.*

Despite the short lifetime in the cell under intense external influences, DBs can affect the properties of the crystal under consideration.

Investigating the dependence of the time of occurrence of discrete breathers on the number of atoms that received an impulse simultaneously, it was found that the probability of the appearance of discrete breathers increases significantly if the percentage of atoms is more than 0.6%. It is obvious that the interaction energy of 1 eV is not enough to excite the DB in one collision, therefore, the larger the number that receives a pulse at a time, the more likely it is to receive a repeated pulse to the aluminum atom, which contributes to the excitation of the DB.

As already discussed, in the electropulse processing of metals, some effects are observed outside the classical theory. Therefore, the inelastic scattering of electrons was simulated. In this case, the interaction of a stream of particles (electrons) with energies up to 2.5 eV with crystal atoms was simulated as follows. Collisions of particles were considered to be fully inelastic, i.e. the electron will completely transfer its energy to the atom upon collision. To simulate the collision of Pt_3Al alloy atoms with electrons during a current pulse, a certain percentage of atoms (from 0.025% to 1.5%) receiving a pulse was randomly selected from the entire array (from 0.025% to 1.5%). The impulses were transmitted to the atoms along one axis corresponding to the <100> crystallographic direction. Thus, the current density passing through the model cell was regulated:

43

$$J = \frac{Ne}{tS},$$ (26)

where N is the number of electrons that passed through the area S in time t, and e is the charge of the electron. By controlling the number of atoms that received an impulse and the time after which a repeated collision occurred, it is possible to estimate the value of the current density, in our case, it varied from 10^4 to 10^6 A/cm^2.

To generate DBs, it is necessary to give the Al atom a velocity of the order of 50 Å/ps. We accept that one collision of an atom with an electron, transfers to it a fifth of this energy or more, i.e. the atom, being stationary, will accelerate by 10 Å/ps or more. Let's calculate what minimum speed an electron should have based on the law of conservation of energy. The energy received by the atom is determined by the formula $E = \frac{m_a v_a^2}{2}$, for one collision it will be about 0.14 eV.

Having carried out a series of computer experiments within the specified ranges of energies of the interaction of electrons with atoms and current densities, it was found that one act of interaction is not enough to excite DBs. An atom must receive successively several collisions with electrons to accumulate the necessary energy for the formation of a discrete breather with a soft type of nonlinearity.

Fig. 36 shows a successive series of experiments with different initial conditions when the interaction energy remained constant at 0.5 eV, and the current density decreased from 10^6 to 10^4 A/cm^2. Obviously, at the maximum current density (Fig. 36a), the probability of obtaining a series of successive collisions leading to energy pumping in the DB is greater. As a result, we can observe (Fig. 36 a) in the computational cell three atoms with an energy of more than 0.8 eV, which are carriers of a nonlinear localized high-amplitude mode, as well as a whole group of atoms of the light sublattice, which can accumulate enough energy during the next collision. By reducing the current density (Fig. 36b) to 10^5 A/cm^2, only one atom with sufficient energy was fixed, and the number of potential discrete breathers decreased significantly. At a current density of 10^5 A/cm^2 or less, it was not possible to detect discrete breathers at the given electron energies. An increase in the interaction energy increased the probability of excitation of discrete breathers at lower current densities as well.

The computer experiments carried out using the molecular dynamics method showed that under intense external influences on the Pt$_3$Al crystal, the generation of discrete breathers with a soft type of nonlinearity at the sites of the Al crystal lattice is possible. Two mechanisms of DB excitation have been identified. In the case of an interaction energy of more than 1.4 eV, the DB excitation can occur in a single collision with a site of the light sublattice of the Pt$_3$Al alloy. If the interaction energy is less than 1.4 eV, then in the case of multiple collisions of particles with atoms of the crystal lattice, energy can be accumulated in the light sublattice of the alloy, thereby creating favorable conditions for the generation of discrete breathers.

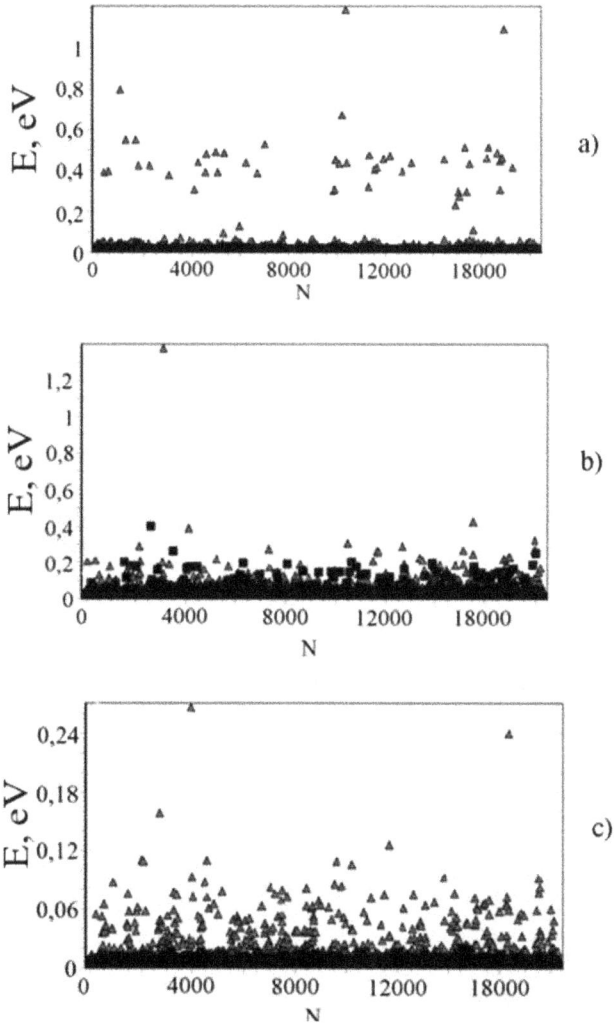

Figure 36: Distribution of energy in the calculated cell of the Pt3Al crystal, along the vertical axis the kinetic energy per atom is plotted, and along the horizontal axis the serial number of the atom in the cell. The triangular marker denotes Al atoms, square marker denotes Pt; (a) at a current density of 10^6 A/cm², (b) at a current density of 10^5 A/cm², (c) at a current density of 10^4 A/cm².

In experimental work on the study of the phonon spectra of NaI, the authors obtained a peak in the bandgap of the phonon spectrum of a crystal in a state of thermodynamic equilibrium at a temperature of 555 K. The results obtained were interpreted as the possible presence of gap discrete breathers in the crystal.

Let us estimate the probability of excitation of two types of discrete breathers in a Pt_3Al crystal. As is known, the probability for atoms and molecules to store a large amount of energy as a result of a series of collisions is negligible. This requires a large number of successive purposeful collisions, as a result of which the atom gains energy, practically without losing it. Therefore, for many processes, only an insignificant fraction of atoms has energy sufficient to overcome the barrier. This share, following the Arrhenius theory, in our case can be determined by the following formula:

$$k = e^{\frac{-E_a}{RT}}, \tag{27}$$

where R is the universal gas constant, T is the crystal temperature, and E_a is the activation energy of the process. It follows from the formula that the fraction of active collisions k depends very strongly on both the activation energy and the temperature.

For a breather with a soft type of nonlinearity, E_a is about $0.8 - 1.0$ eV. Based on these data Fig. 37 shows the dependence of the value of k on the temperature of the crystal. From the data obtained, it can be concluded that up to 0.01% of the atoms of the model crystal at a temperature of 1000 K can be DB carriers.

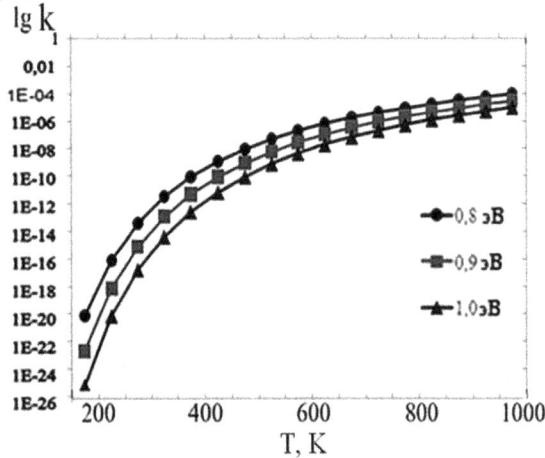

Figure 37: *Estimation of the probability of excitation of a breather with a soft type of nonlinearity with different energies depending on the crystal temperature (along the abscissa is the value of the temperature T in K, along the ordinate, is the decimal logarithm of k).*

In turn, a breather with a hard type of nonlinearity has a higher activation energy than a DB with a soft one. The minimum energy at which such a breather can stably exist is 1.8 eV. In this case, the probability of its excitation in the state of thermodynamic equilibrium at a temperature of 600 K is $7.6 \cdot 10^{-19}$. Taking into account that we expended more energy, 6 eV, to excite the DB, then the probability of the appearance of such objects in the crystal drops sharply to the value $k = 4 \cdot 10^{-51}$. The obtained values indicate that the probability of excitation of discrete breathers with a hard type of nonlinearity in a state of thermodynamic equilibrium is extremely small, and their spontaneous excitation without external influences on the crystal is not possible.

2.6 Discrete breathers near the surface of a Pt$_3$Al crystal model with an EAM potential

Considering discrete breathers in various crystals, most often we are talking about their properties in "ideal" lattices, without any structural defects. However, it is also obvious that defects and various inhomogeneities of the medium affect the characteristics of such objects.

This section is devoted to the study of discrete breathers with a soft type of nonlinearity near the surface of a crystal of stoichiometric composition A$_3$B, by the example of Pt$_3$Al using the EAM potential. As shown above, discrete breathers can exist in this crystal, while discrete breathers with a soft type of nonlinearity can be excited spontaneously at high temperatures, or when the crystal is periodically exposed to frequencies close to the frequencies of discrete breathers, in addition, the possibility of excitation of such objects by a stream of high-energy particles is shown. Thus, the study of discrete breathers near the surface seems relevant and useful when searching for them in real experiments.

It is worth remarking on the Pt$_3$Al crystal model. Pair potentials do not always correctly describe the elastic properties of crystals, overestimate the formation of a vacancy in a metal, and do not describe the decrease in the bond energy per bond with an increase in the coordination number, so their use for describing the surface is unacceptable. When conducting a study of discrete breathers near the surface, the potential obtained by the embedded atom method was used.

The model under consideration was a bulk fcc Pt$_3$Al crystal with the L1$_2$ superstructure. The number of particles in the considered models varied from 108,000 particles to 191,500 particles, depending on the orientation of the crystallographic planes.

For this model, the phonon spectrum of the crystal was calculated. The LAMMPS software package was used in the calculations, which includes the procedures necessary for these purposes, based on the Fourier transform of the autocorrelation functions of atomic displacements versus time.

And also the dependence of the frequency on the amplitude for a breather with a soft type of nonlinearity in the bulk of the crystal was obtained (see Fig. 38). The model preparation process consisted of the initial relaxation of the crystal with free boundary conditions along all axes at temperatures of 500 K followed by cooling to 0 K, which made it possible to eliminate the effect of thermal vibrations on the breathers in the crystal.

Three planes were considered in such a way that they emerged perpendicular to the crystal surface, (100), (110), and (111). Further, in all cases, free boundary conditions were imposed along the Y axis and periodic ones along the Z and X axes.

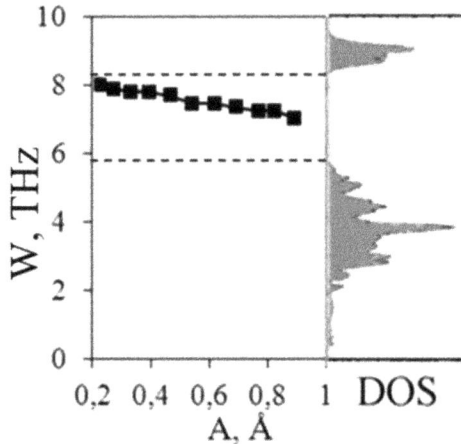

Figure 38: The dependence of the frequency of oscillations of discrete breathers in the bulk of a crystal on the amplitude is shown in comparison with the density of phonon states of a crystal with an EAM potential.

For the (100) plane, two variants of the surface are possible: (1) the upper layer consists of Pt atoms (Fig. 39 a), (2) the upper layer consists of Al atoms (Fig. 39 b).

When the atom numbered 1 in Fig. 39a is deflected along the vertical axis Y, after a few picoseconds, it transfers its energy to atom number 2, on which a discrete breather is subsequently formed with characteristics corresponding to the breather breath in the volume of the crystal, i.e. the surface subsequently does not affect it, because such discrete breathers have a high degree of spatial localization.

When excited along the horizontal X-axis and perpendicular to the pattern plane, the DB remained stable for a long time and its characteristics also corresponded to the DB in the volume of the crystal.

In the case when the surface layer was an aluminum layer (see Fig. 40b), stable vibrations were possible only perpendicular to the surface along the Y axis, shown by arrows in the figure. Attempts to excite DBs in other directions led to the dissipation of energy over the crystal in the form of thermal vibrations of atoms. This is due to the formation of a characteristic surface relief from the difference in the binding energy of the various components of the alloy.

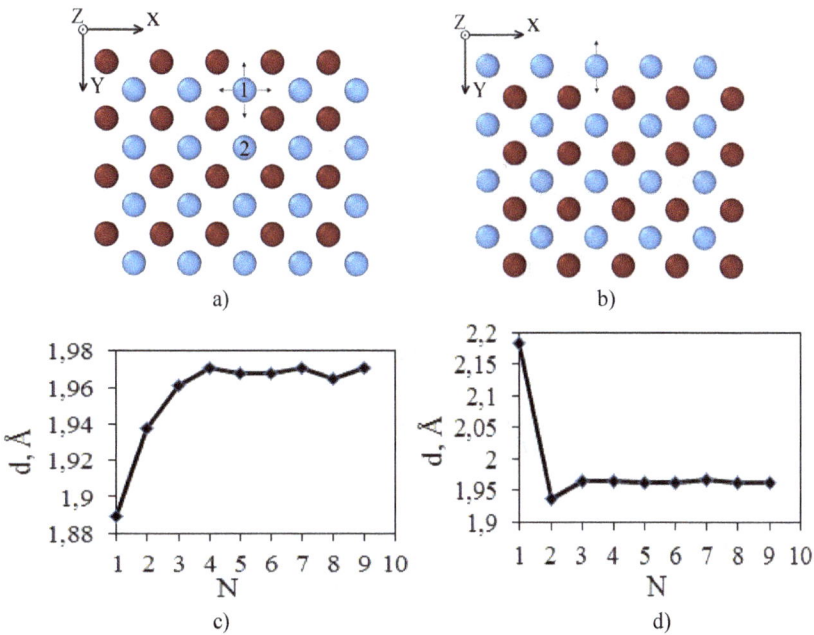

Figure 39: *Visualization of one atomic plane (100) of a Pt₃Al crystal, the X axis is directed along <010>, the Y axis is <001>, Z is <100>; (a) the surface layer is Pt, (b) the surface layer is Al, the arrows show the possible directions of stable vibrations discrete breathers, (c) and (d) show the dependence of the distance between adjacent atomic rows from the surface into the depth, N is the serial number of a pair of neighboring atomic layers from the surface into the depth of the crystal.*

When considering the (110) plane, three options for the arrangement of atoms near the surface are possible (Fig. 41).

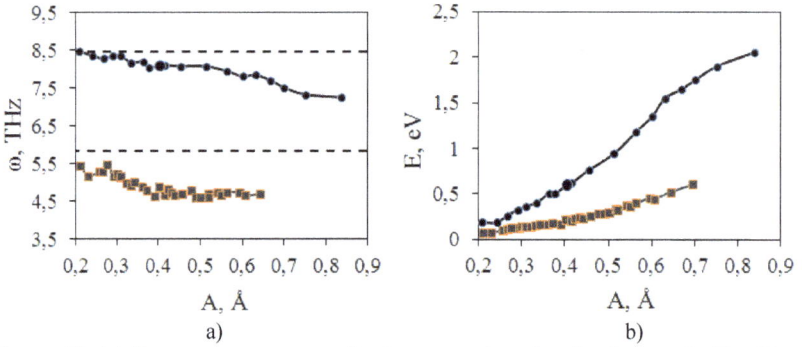

Figure 40: *(a) Frequency versus amplitude: a round marker for the case in Fig. 39a, a square marker for the case in Fig. 3.39b, (b) dependence of the energy localized on the DB on the oscillation amplitude: a round marker for the case in Fig. 39a, a square mark for the case in Fig. 3.39b.*

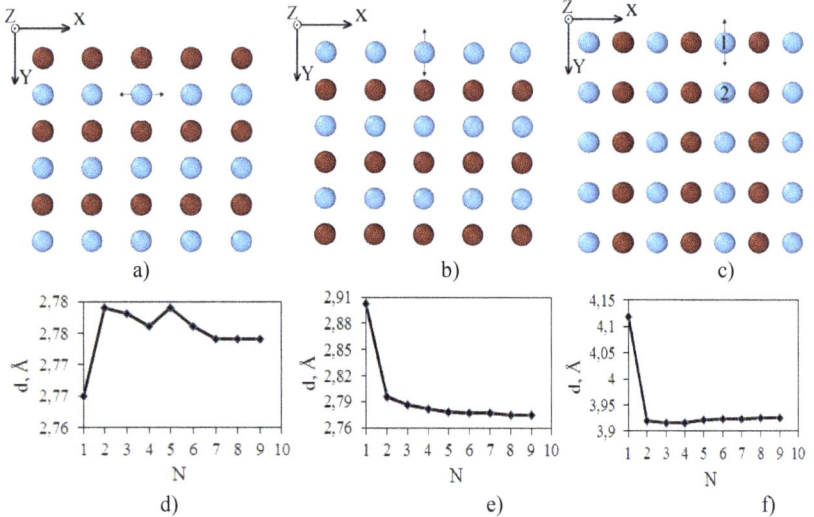

Figure 41: *Visualization of one atomic plane (110) of a Pt₃Al crystal; (a) the X axis is directed along <001>, the Y axis is <11 0̄>, Z is <110>, (b) the X axis is directed along <001>, the Y axis is <11 0̄ >Z – <110>, (c) X-axis is directed along <11 0̄>, Y-axis - <001>, Z – <110>, arrows show possible directions of stable oscillations of DB, (d), (e) and (f) show the dependence of the distance between adjacent atomic rows from the surface into the depth, N is the serial number of a pair of neighboring atomic layers from the surface into the depth of the crystal.*

In the case in Fig. 41a, stable oscillations along the X-axis are possible, while the characteristics of the discrete breathers corresponded to the discrete breathers in the volume of the crystal. In the case of Fig. 41b, the surface layer is Al, and vibrations were only possible perpendicular to the surface. Fig. 41c shows that when atom number 1 deviates along the vertical Y axis, it transfers its energy to atom number 2 after a few picoseconds.

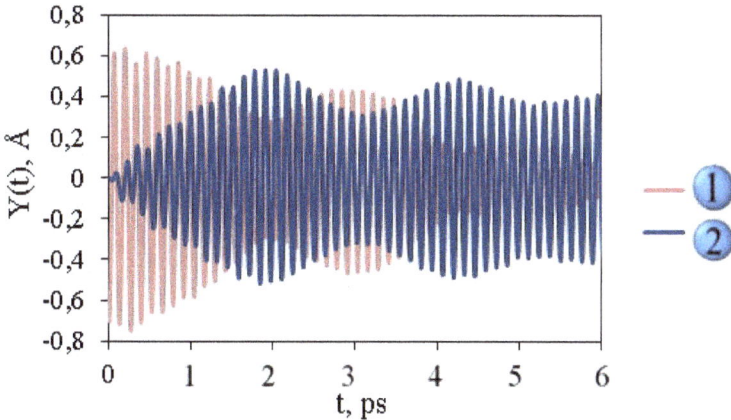

Figure 42: *Transfer of energy from atom 1 to atom 2 (Fig. 41c).*

Considering the (111) plane, we chose the options when Al atoms are present on the surface. Other configurations correspond to the cases when Al atoms are several layers away from the surface.

In Fig. 43a, stable oscillations are possible both along the X axis and along the Y axis, however, in the case of initiation of oscillations along the X axis, the oscillations are reoriented along the Y axis (Fig. 44).

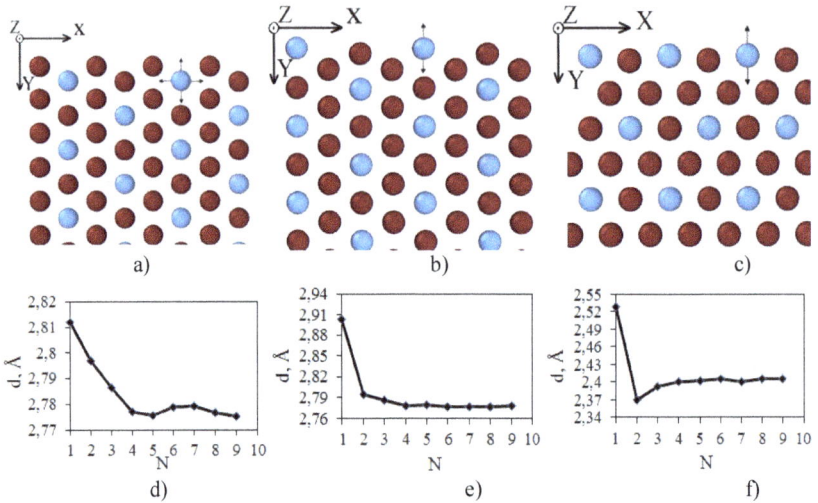

Figure 43: *Visualization of one atomic plane (111) of a Pt3Al crystal; (a) the X axis is directed along <112‾>, the Y axis is <11‾0>, Z is <111>, (b) the X axis is directed along <112‾>, the axis Y - <11‾0>, Z - <111>, c) X-axis is directed along <11‾0>, Y-axis - <112‾>, Z - <111>, (d), (e) and (f) show the dependence of the distance between adjacent atomic rows from the surface into the depth, N is the serial number of a pair of neighboring atomic layers from the surface into the depth of the crystal.*

This graph makes it possible to notice that there is a displacement of the center of vibrations of the atoms of the atom on which the DB is localized.

In Figs. 43b and 43c stable oscillations are possible only for the cases shown by the arrows, i.e. perpendicular to the crystal surface.

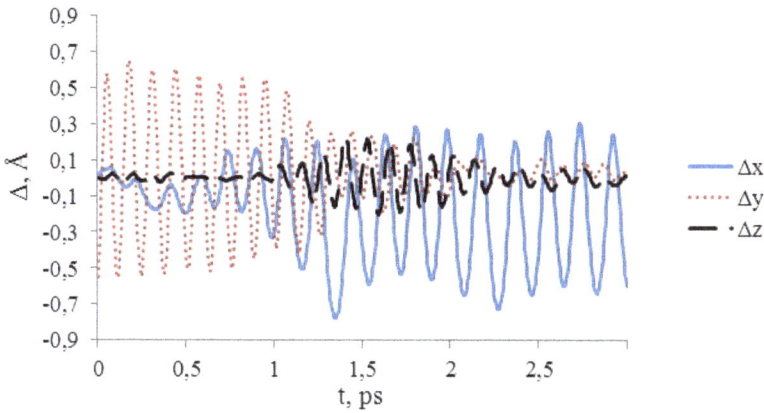

Figure 44: *The process of changing the polarization of atomic vibrations in Fig. 43a.*

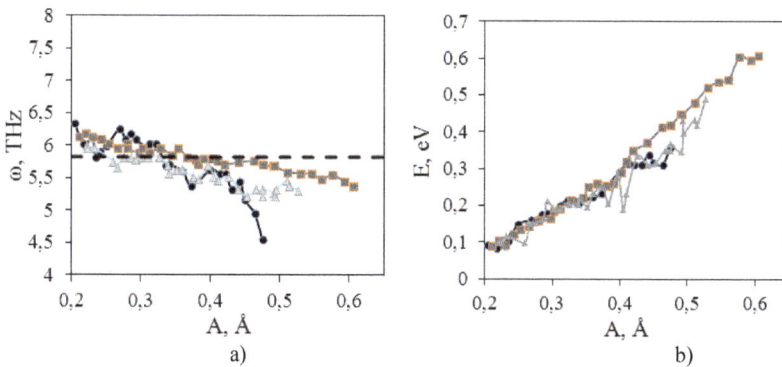

Figure 45: *(a) Oscillation frequency versus amplitude: round mark for the case in Fig. 43a, a square mark for the case in Fig. 43b, the triangular mark for the case in Fig. 43c, (b) dependence of the energy localized on the DB, round mark for the case in Fig. 43a, a square mark for the case in Fig. 43b, the triangular mark for the case in Fig. 43c.*

Thus, if the breather is located near the surface, vibrations are possible mainly perpendicular to the surface, while the directions of vibrations may not correspond to the case in the volume.

The nucleus of a discrete breather can migrate to the neighboring atom of the light sublattice deep into the crystal. The value of vibration amplitudes in most cases for stable vibrations perpendicular to the surface does not exceed 0.5-0.6 Å, while stable vibrations up to an amplitude of 0.9 Å are possible in the volume. The amount of energy stored on the discrete breathers near the surface does not exceed 0.6 eV; discrete breathers with a soft type of nonlinearity up to 2.5 eV are possible in the bulk of the crystal.

The energy of discrete breathers on the surface is significantly (three to four times) less than the energy in the bulk of the crystal. The results obtained show the possibility of energy localization on the surface of crystals, which may be important for surface physics.

2.7 On the possibility of the existence of discrete breathers based on the quantum mechanical model of the Pt₃Al crystal

The question related to the existence of such localized objects as discrete breathers in real crystals is still open, due to the problematic nature of their observation in real computer experiments, as well as the complexity of the calculation from first principles for three-dimensional crystals. In this section, the main characteristics of the Pt₃Al system are calculated: dispersion curves, the density of phonon states, and the distribution of the electron density near the Al atom. These parameters make it possible to estimate the possibility of energy localization in the form of a discrete breather with a soft type of nonlinearity in such a crystal.

Fig. 46 shows the dispersion curves and the density of phonon states; in contrast to the above examples, there is no clear gap in the phonon spectrum of the crystal.

However, this result does not indicate the absence of conditions for the existence of such objects along certain directions in the crystal, such as [100].

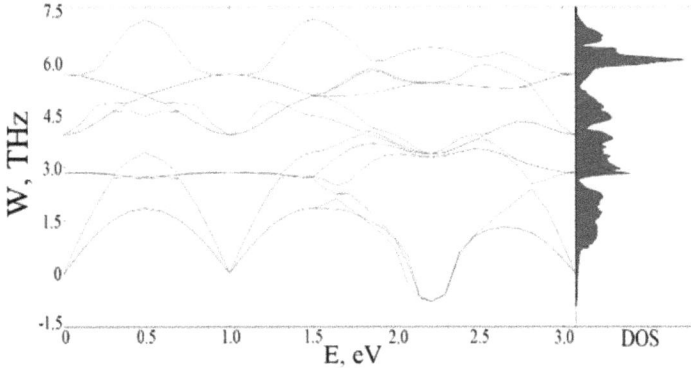

Figure 46: *Dispersion curves and distribution of the density of phonon states of a Pt₃Al crystal.*

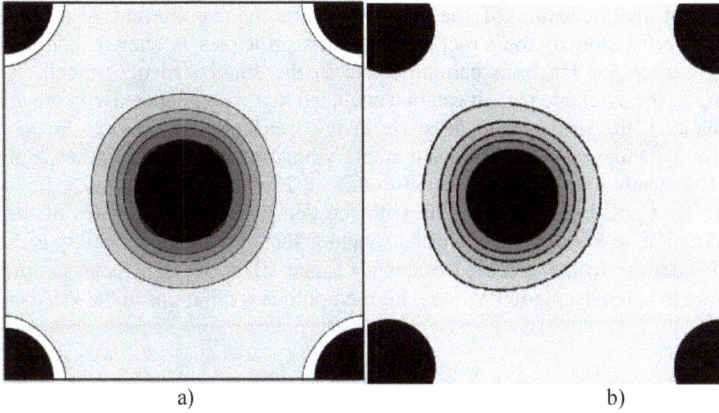

a)	b)

Figure 47: *Distribution of the electron density near the Al atom (a) in the equilibrium position (b) with a shift of 0.5 Å along the [100] direction.*

The results presented will be further calculated in dynamics, which will allow a more complete picture of the behavior of the Al atom at sufficiently large deviations from the equilibrium position. Fig. 47 shows the electron density distribution near the Al atom when the atom is deflected and in the equilibrium position.

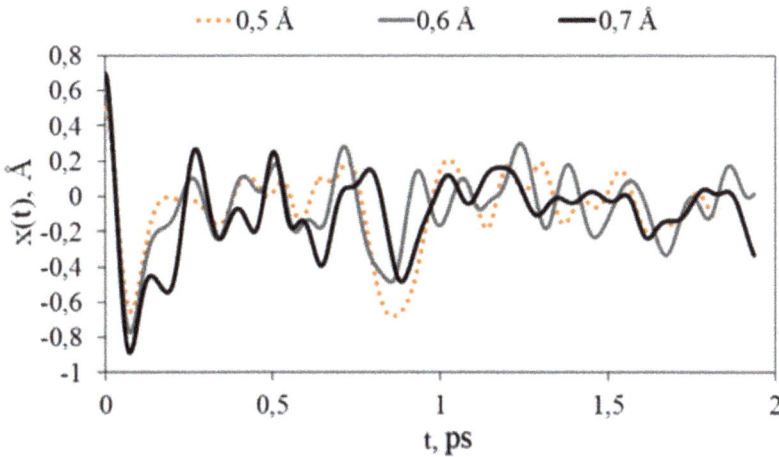

Figure 48: *Dependence of the coordinate of the Al atom relative to the equilibrium position for the initial amplitudes: 0.5, 0.6, and 0.7 Å.*

Modeling of the deviation of the Al atom from the equilibrium position and the subsequent relaxation of the structure from first principles is shown in Fig. 48. The resulting curves are far from harmonic due to the small size of the cell. However, according to the available data, it can be concluded that with an increase in the amplitude of oscillations, the tendency to decrease in the frequency remains, as in the case of molecular dynamics models. The most stable vibrations were at an initial amplitude of 0.6 Å. The steady amplitude was equal to 0.2 - 0.25 Å and the frequency 5.5 - 5 THz, respectively. Comparing these results with the density of phonon states of the crystal obtained by the *ab initio* method, we can conclude that the vibration stability is due to the lowest PS density in the specified frequency range. Thus, we can speak of sufficiently long (more than ten oscillation periods) high-amplitude oscillations in the Pt_3Al crystal.

Chapter 3. Discrete breathers in CuAu and CuPt₇ crystals

3.1 Search for discrete breathers in a CuAu crystal

We considered the CuAu biatomic system in order to identify various types of discrete breathers in this crystal. The method of molecular dynamics was chosen to deal with the problem, as it has proven itself in the study of such processes. This method was implemented using the LAMMPS molecular dynamics simulation package, which uses well-proven multi-particle interatomic potentials built by the embedded atom method. The EAM potential for the CuAu system was used in the calculations.

To analyze the possibility of the existence of discrete breathers in the CuAu crystal, we calculated the density of phonon states of the crystal under consideration (see Fig. 49). The absence of a gap in the phonon spectrum of CuAu indicates that it is unlikely to contain discrete breathers with a soft type of nonlinearity.

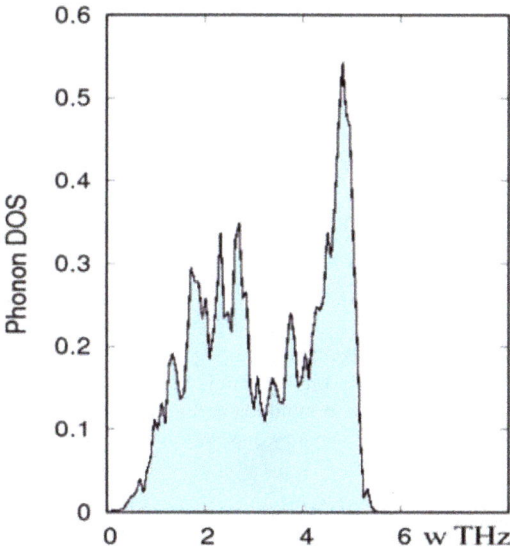

Figure 49: The density of phonon states CuAu crystal.

The process of searching for discrete breathers in crystals involves the selection of initial conditions - the deviations of atoms from an equilibrium position or the setting of initial velocities. In the case of a large mass difference in biatomic crystals (more than 4 times), it is sufficient to remove one or two atoms from the equilibrium position in a certain direction, which leads to the formation of discrete breathers with a mild type of nonlinearity. Such an approach did not lead to the formation of discrete breathers along any of the crystallographic directions due to the characteristic phonon spectrum under normal conditions in the studied crystal.

For pure metals or alloys with a smaller spread of component masses, the conditions for the excitation of discrete breathers with a hard type of nonlinearity are more specific. Thus, Dmitriev S.V. proposed an ansatz for the excitation of discrete breathers in pure fcc and bcc metals, which implies setting the breather profile by imparting displacements to them using special functions. Displacements of atoms were carried out in such a way that neighboring atoms carried out vibrations in the antiphase.

In this work, to create the initial DB profile, we used the Gaussian function (28) adapted for the crystal conditions

$$f(x) = A_0 \ e^{-\frac{(x-b)^2}{2c^2}} \tag{28}$$

The parameter b was excluded from this equation because an ideal crystal without topological defects was considered, thus, we have a function of the form:

$$f(x) = A_0 e^{-\frac{x^2}{2c^2}} \tag{29}$$

where A_0 specifies the initial amplitude of the central atoms of the discrete breather, x is the relative coordinate of a pair of atoms in the row, and parameter C is the degree of spatial localization of the discrete breather. By varying the value of A_0 and C, we select the profile of a discrete breather, thereby setting the initial deviations from the equilibrium position for the atoms entering the DB (see Fig. 50).

Considering the CuAu crystal for the possibility of forming discrete breathers with a hard type of nonlinearity, we checked the close-packed directions for the Cu and Au sublattices. A discrete breather was obtained along the [110] direction for copper atoms. The existence of discrete breathers on Au atoms is not possible, since their deviation from the equilibrium position leads to rapid excitation of neighboring atoms of the lighter Cu sublattice, which in turn leads to energy dissipation in the crystal.

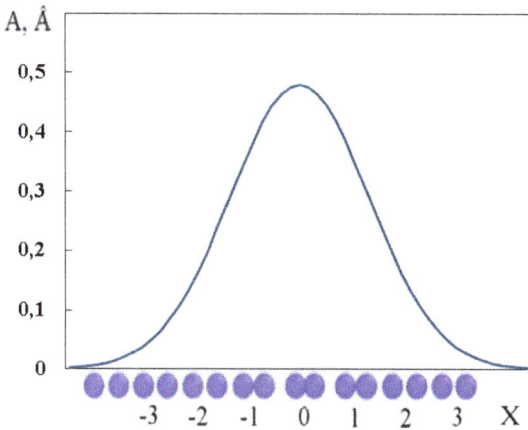

Figure 50: *Basic profile discrete breather given by the function (2) for a number of coppers along the [110].*

The longest vibrations were obtained for the parameters of equation (29) $A_0 = 0.48$ Å, $C = 0.75$. The DB lifetime in this case exceeded 50 ps. The formed discrete breather was localized on six to eight copper atoms vibrating in antiphase.

The most important characteristic of a discrete breather is the dependence of its frequency on the amplitude of atomic vibrations. By varying A_0 in equation (29) and taking the results from the model, the corresponding curve was obtained, shown in Fig. 51. The oscillation frequency was fixed after 20–30 oscillations from the beginning of the experiment. As can be seen from Fig. 51, with an increase in the initial amplitude of more than 0.55 Å, there is practically no change in frequency. This can be explained by the fact that a significant deviation of atoms from the equilibrium position leads to the excitation of a heavy sublattice of the crystal, through which part of the energy is dissipated into the phonon subsystem of the latter. A further increase in the initial amplitude led to the complete dissipation of energy over the crystal without the formation of a discrete breather.

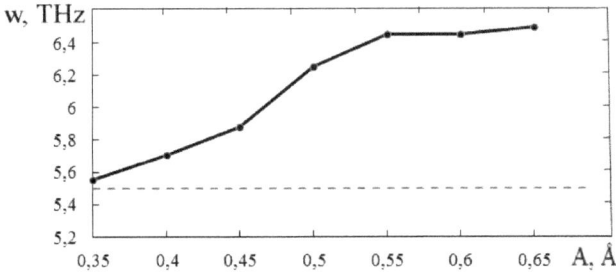

Figure 51: *The frequency ω vibrations of the atoms belonging to the discrete breathers, the amplitude A of the central atoms DB; the dashed line shows the upper limit of the phonon spectrum of the crystal CuAu.*

An equally important value is the energy that the DB can localize on itself for a long time. Fig. 52 shows the dependence of the energy of a discrete breather with a hard type of nonlinearity on the oscillation amplitude of the central DB atoms, which it can maintain for at least 5 ps.

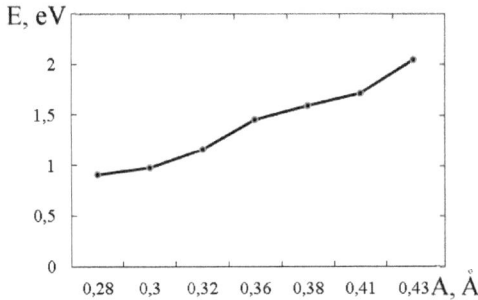

Figure. 52: *The dependence of the energy of the discrete breather with a hard type of the nonlinearity of the amplitude of oscillation of the central atoms.*

The degree of spatial localization of the DB during its existence can be estimated from the energy profile shown in Fig. 53. A discrete breather is localized on 6-8 Cu atoms and can collectively focus up to 2.1 eV.

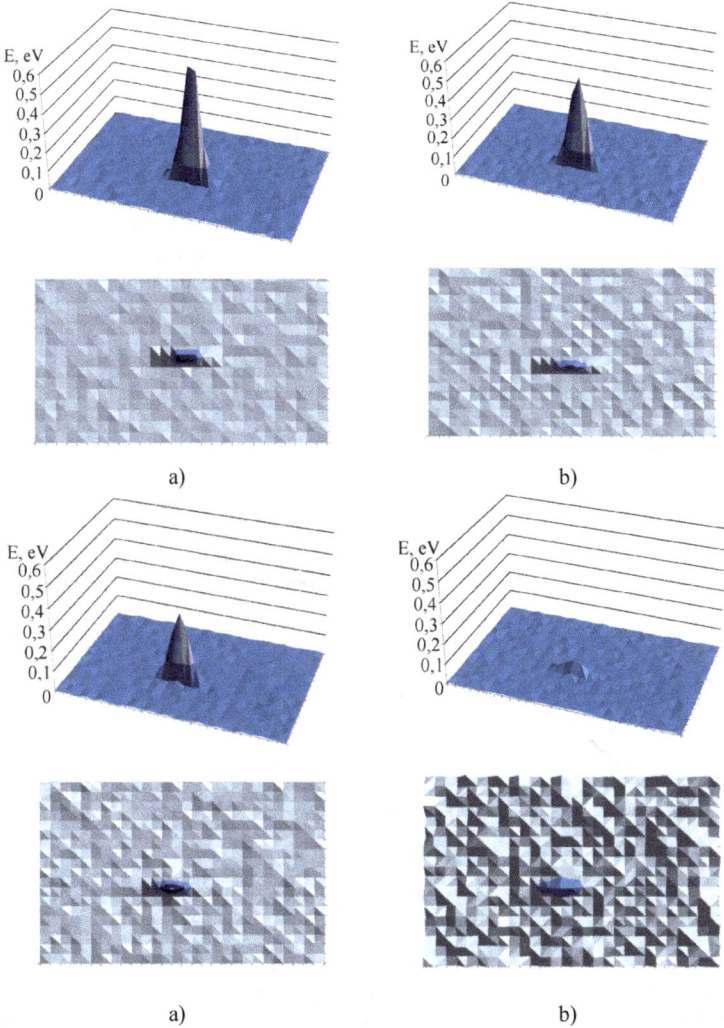

Figure 53: *Evolution of the energy profile of the discrete breather with a hard type of non-linearity: a) Under the instant 5 pc from the beginning of the experiment, the total energy of 1.71 eV DB, b) As at time of 30 ps from the start of the experiment, the total energy of 1.45 eV DB, c) at the time of 45 ps from the start of the experiment, the total energy of 0.98 eV DB, d) into the time of 50 ps time from the beginning of the experiment, the total energy of 0.25 eV DB.*

A discrete breather with a soft type of nonlinearity can be obtained in crystals by deforming them or by underestimating the mass of the alloy component, which led to the formation of a gap in the phonon spectrum of the crystal, thereby providing the necessary conditions for the existence of discrete breathers.

We deformed the CuAu crystal taking into account the Poisson principle, i.e. the volume of the model cell was preserved. A characteristic indicator for crystals with a tetragonal structure is the ratio of the lattice parameters c/a, for CuAu under normal conditions this ratio is 0.92. Deforming the crystal along the crystallographic direction [001] corresponding to the lattice parameter c, taking into account changes in the lattice parameter a to preserve the volume of the model under consideration, we obtained the phonon density of states of the crystal.

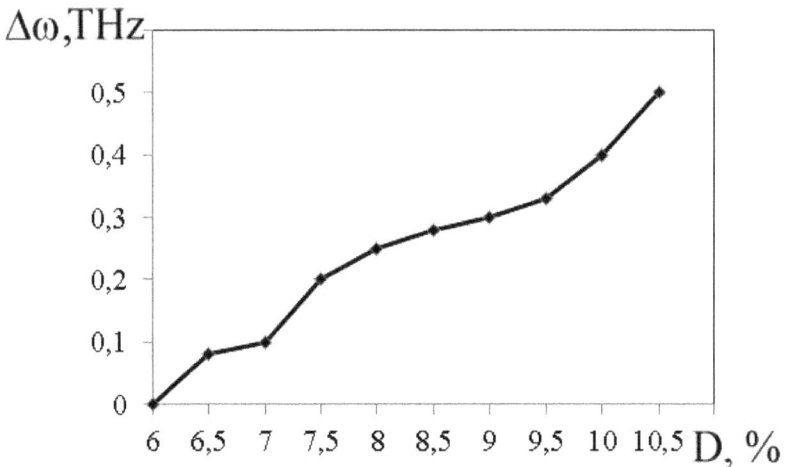

Figure 54: *Dependence of the gap width of the phonon spectrum of the crystal CuAu the magnitude of deformation D along the [001] direction.*

A fairly wide gap in the phonon spectrum of the crystal was obtained at $c/a = 0.8$, which amounted to 9.5% of the compression strain along the [001] direction (see Fig. 54).
A further increase in the compression strain led to a widening of the gap, but at the same time had a negative effect on the stability of the model.

Under tensile deformation of the CuAu crystal, i.e. as the c/a ratio increased, no gap appeared in the phonon spectrum of the crystal.

To search for a discrete breather with a soft type of nonlinearity, one of the Cu atoms was deviated from the equilibrium position along different crystallographic directions. As expected, in this case we managed to obtain a discrete breather along the [100] direction.

The frequency of the obtained discrete breather with a soft type of nonlinearity lies in the gap of the phonon spectrum and corresponds to 3.5 THz. The lifetime of the obtained localized oscillations is about 2 ps or more than 15 oscillation periods. This type of discrete breather is localized mainly on one Cu atom. Due to the narrow gap in the phonon spectrum, it is not possible to obtain the dependence of the frequency on the amplitude in a wide range.

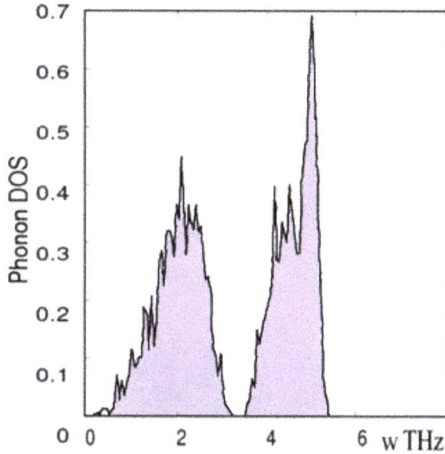

Figure 55: *The density of phonon states of the crystal CuAu, with c/a = 0.8 (compression deformation was 9.5%).*

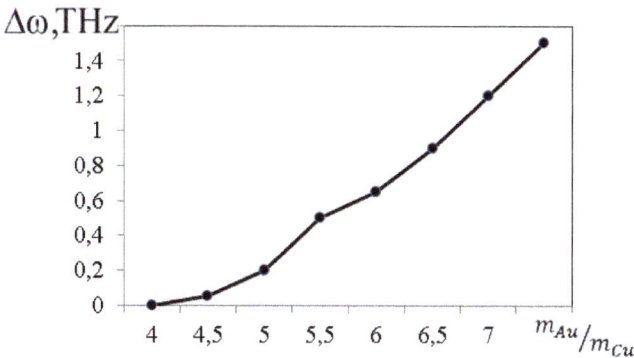

Figure 56: *Dependence of the gap width of the phonon spectrum of the crystal CuAu on the ratio of the masses Au alloy component to the Cu.*

Further, experiments were carried out with a reduced mass of copper without deformation. A similar gap in the FS, as in Fig. 55, was achieved by reducing the mass of Cu to 36.0 a.e.m. (normal 63.546 a.e.m.) or it corresponded to a gold to copper mass ratio of 5.5 (see Fig. 56). The duration of localized oscillations, in this case, was similar, about 15 oscillation periods, which is apparently due to the geometry of the crystal.

Compressive deformation, together with a decrease in the mass of copper to 20 a.e.m. led to the formation of a wider gap in the FS with a stable crystal lattice, which made it possible to increase the lifetime of DBs with a soft type of nonlinearity up to 100 oscillation periods.

For this case, the energy localized on a discrete breather with a soft type of nonlinearity was obtained. The degree of spatial localization and the energy profile of DBs with a soft type of nonlinearity are shown in Fig. 57.

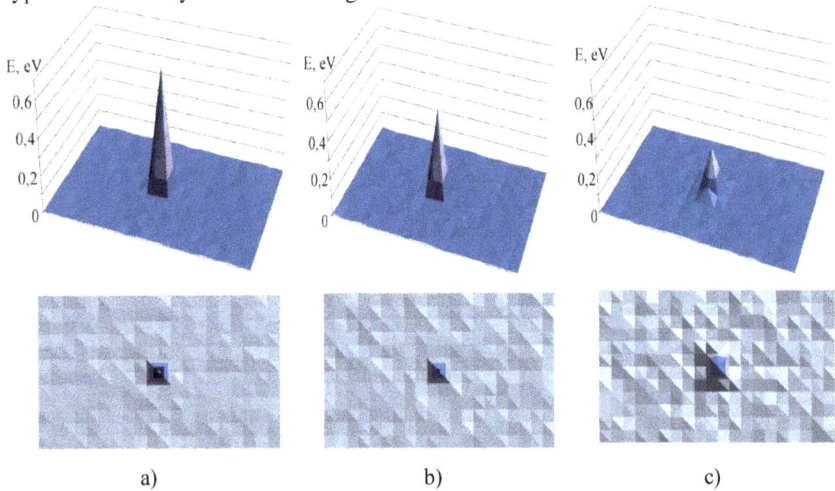

a)   b)   c)

Figure 57: *The evolution of the energy profile of the discrete breather with a mild type of the nonlinearity: a) under point in time of 1 ps from the beginning of the experiment, the total energy of 0.72 eV DB, b) at time of 2.5 ps from the beginning of the experiment, the total energy of 0.53 eV DB, c) at time 7 ps from the beginning of the experiment, the total energy of 0.28 eV DB.*

The results obtained indicate that the discrete breather with a soft type of nonlinearity is localized mainly on one copper atom. Despite this, it can collectively concentrate the energy of the order of 0.9 eV, while a DB with a hard type of nonlinearity is mainly localized on six atoms and their total energy is up to 2.1 eV.

Thus, the molecular dynamics method has demonstrated the possibility of excitation of a discrete breather with a hard type of nonlinearity in a CuAu crystal on a copper sublattice. Its dependence of the frequency on the amplitude is obtained, and the optimal parameters of the function are selected, which describe the initial profile of the discrete breather. The spatial localization is considered, and the energy profile of the breather is presented, which makes it possible to estimate the value of the DB energy. It has been found that a gap can form in the phonon spectrum of a CuAu crystal and a discrete breather with a soft type of nonlinearity on Cu atoms can exist in it under compressive deformation. The dependence of the gap width on the applied strain along the [001] crystallographic direction is obtained. It is shown that an artificial reduction of the mass of copper atoms also leads to a gap in the phonon spectrum and the creation of conditions for the existence of a discrete breather with a soft type of nonlinearity localized along the [100] direction. Upon deformation or reduction of the mass, the DB lifetime is about several tens of oscillation periods; to increase the duration of oscillations, deformation was applied to the crystal and the mass of copper atoms was reduced together. For this case, the DB energies are obtained, the evolution of the DB energy profile and its spatial localization are presented.

3.2 Dynamics of a discrete breather with a hard type of nonlinearity in a CuAu crystal

We considered the CuAu biatomic system in order to reveal the possibility of motion of a discrete breather along close-packed directions of the crystal. To create the initial DB profile, the bell-shaped Gaussian function (29) was used.

To specify a moving DB, it is necessary to introduce asymmetry into the discrete breather profile. For this purpose, a factor γ at C was introduced to redefine one of the branches of function (30), and the second branch was calculated at $\gamma = 1$:

$$f(x) = A_0 e^{-\frac{x^2}{2\gamma \cdots C^2}} \tag{30}$$

Figs. 58 and 59 show the initial profile of the moving ($\gamma <> 1$) and stationary ($\gamma = 1$) discrete breathers. By selecting the multiplier for the right branch, one can obtain DBs with different initial speeds of movement along the crystal.

The results obtained indicate that the rate of such displacements of discrete breathers is much lower than the speed of sound in the crystal under consideration. The movement of a discrete breather with an asymmetric profile occurs towards a steeper branch of the function. A DB can overcome several tens of interatomic distances in a crystal along a close-packed row during its lifetime. As the initial parameters for the moving breather, we chose the parameters of equation (29) at which the DB existed in the stationary state for the maximum amount of time. The meaning of these parameters is as follows: $A_0 = 0.48$ and $C = 0.75$.

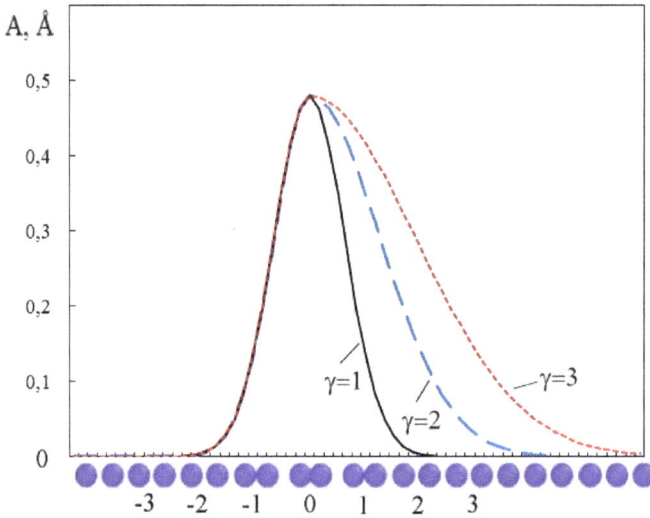

Figure 58: *Specifying a discrete breather profile using (29) and (30).*

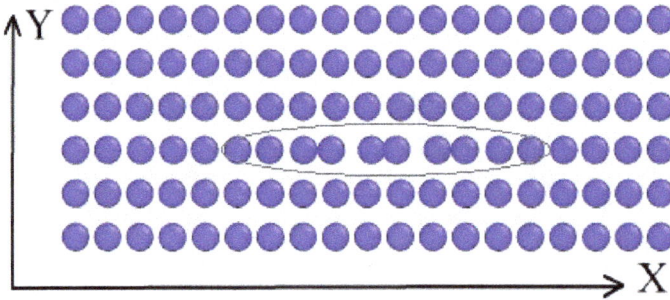

Figure 59: *Visualization of displacements of atoms in the plane of Cu atoms, the X axis is directed along the <110> crystallographic direction, while the Y axis is along the <001>.*

Next, the parameter γ was varied for the right side of the DB profile. As the main parameter that was tracked, the distance traveled by the DB during its lifetime was chosen. The graph of the obtained dependence is shown in Fig. 50.

The results indicate a weak mobility of such objects in the crystal under consideration. However, energy transport along close-packed directions of the Cu sublattice is still possible.

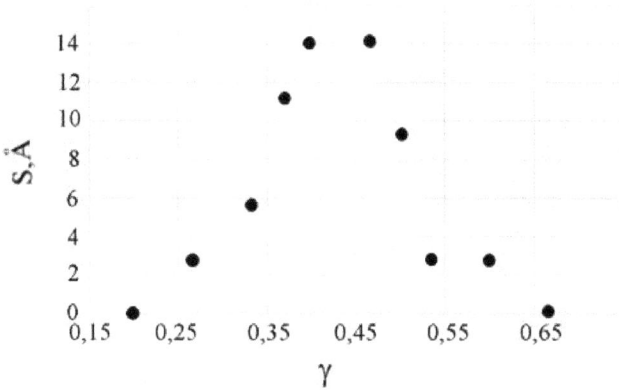

Figure 60: *Dependence of the distance traveled S by the breather on the parameter* γ.

Thus, the possibility of excitation of a movable discrete breather with a hard type of nonlinearity in a CuAu crystal on a copper sublattice has been demonstrated. Its main characteristics are analyzed, the possibility of its movement in a crystal is estimated, as well as the influence of the initial conditions on the lifetime of a moving discrete breather.

3.3 High-amplitude excitations of the $CuPt_7$ crystal lattice

The models we are considering is a bulk fcc $CuPt_7$ crystal containing 23328 particles (Fig. 61) interacting via the embedded atom method (EAM potential). Two variants of a lattice with orthogonal basis vectors (Fig. 61a) and a trigonal configuration (Fig. 61b) were considered.

To calculate the density of phonon states, we also used the LAMMPS software package, which includes the procedures necessary for these purposes, based on the Fourier transform of the autocorrelation functions of atomic displacements versus time. The results for a lattice with orthogonal basis vectors are shown in Fig. 61c, for the trigonal structure in Fig. 61d.

An analysis of the crystal structure and the calculated phonon spectrum suggests the possible existence of such objects as a discrete breather with a soft type of nonlinearity in $CuPt_7$. For example, in the crystal shown in Fig. 15b, there is a band gap, which is a necessary condition for the existence of a discrete breather. In addition, the lighter copper atoms are surrounded by heavy platinum atoms, which should also have a favorable effect on the localization of light sublattice vibrations. However, the mass of copper is less than three times less than the mass of platinum, which is often insufficient to form the conditions for the existence of a discrete breather.

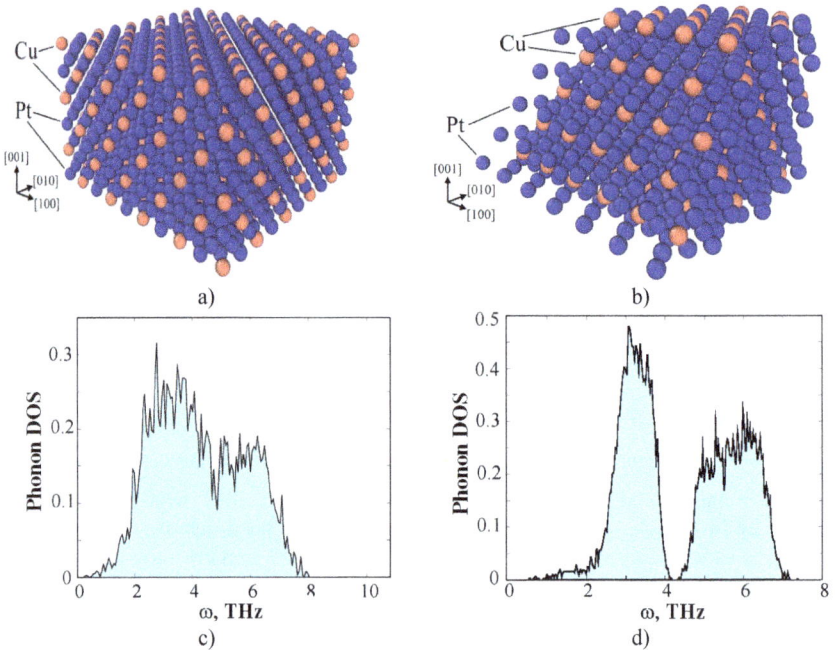

Figure 61: *CuPt₇ crystal structure, (a) with orthogonal basis vectors, (b) trigonal structure, (c), (d) corresponding phonon density of states of crystal models.*

Further, the dependence of the oscillation frequency of Cu atoms on the amplitude was obtained for both the above types of crystal syngonies. As an example, consider vibrations along the <001> direction.

The results obtained in Fig. 62 speak of a hard type of nonlinearity of vibrations for both systems of the $CuPt_7$ crystal, which does not allow the vibrational frequency of the copper atom to fall into the band gap of the phonon spectrum for the trigonal system of the crystal. Thus, the possibility of excitation of discrete breathers with a soft type of nonlinearity is excluded. It should be noted that at sufficiently large amplitudes, for these crystal structures, long but damped vibrations of Cu atoms were observed. This was most clearly manifested in a cubic crystal, due to the higher order of symmetry. In this case, one can speak not of discrete breathers, but of nonlinear high-amplitude oscillations of lattice sites in a defect-free $CuPt_7$ crystal.

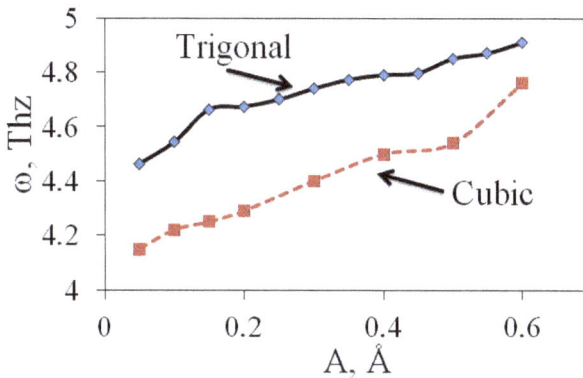

Figure 62: *Vibration frequency of the Cu atom in a CuPt7 crystal as a function of the amplitude for the trigonal and cubic systems of the CuPt7 lattices.*

Below, using the example of a cubic crystal, we consider the features of such nonlinear modes in CuPt7. The frequencies of their vibrations lie in the spectrum of the crystal, so the energy dissipates quickly enough to neighboring atoms and further throughout the crystal. The most stable vibrations of the copper atom were obtained for initial amplitudes of 0.96 Å. For comparison, this dependence is also shown for an initial amplitude of 0.3 Å. In this case, the oscillations decay much faster and then do not recover.

Thus, it is shown that the excitation of breathers with a soft type of nonlinearity is not possible in this crystal. In this case, at certain initial amplitudes, the lifetime of individual modes can significantly exceed the average lifetime of high-amplitude thermal vibrations of atoms. The carriers of such modes are Cu atoms; this is most pronounced for a crystal with a cubic system due to a higher order of symmetry.

Chapter 4. A statistical approach to the analysis of discrete breathers

4.1 Quasi-breathers in Pt$_3$Al

A discrete breather, as an object strictly periodic in time, is obtained in numerical simulation only in the case of an ideal adjustment of the initial conditions of the Cauchy problem to a certain low-dimensional manifold in a multidimensional space of all possible initial values of the coordinates of individual particles and their velocities. Such fine-tuning is difficult to implement even in a computational experiment. Moreover, it is practically impossible to do this when setting up any physical experiments, especially in cases where breather-like objects arise spontaneously.

As already noted, in the work of G.M Chechin. the concept of quasi-breathers was put forward, as some dynamic objects localized in space, but not strictly periodic in time. At the same time, a certain criterion for the proximity of a quasi-breather to the corresponding exact breather was formulated, based on the calculation of the root-mean-square deviation $\eta(t_k)$ of the oscillation frequencies of individual breather particles found on a certain interval in the vicinity of the moment t_k, and the calculation of the root-mean-square deviation of the oscillation frequencies of the selected j-th breather particles at different time intervals.

Thus, all the objects we call, are quasi-breathers. In this section, a statistical evaluation of the characteristics of quasi-breathers in model lattices of composition A$_3$B is carried out.

Unlike exact discrete breathers, quasi-breathers are not strictly time-periodic dynamic objects, although they are localized in space. They arise for any sufficiently small deviations from the exact breather solutions in the multidimensional space of all possible initial conditions when solving the Cauchy problem for the original differential equations, since in this case there is no complete suppression of the contributions from oscillations of peripheral particles with their own frequencies. Thus, the "weakening of the dictatorship" on the part of the breather core (in the case of the symmetric breather we are considering, the core is also formed by one central particle, and in the case of an antisymmetric breather, two of its central particles) leads to the presence in the breather solution of small contributions having different frequencies. These small contributions can be found in the vibrations of all particles in the chain, in particular, the central ones. If we find sufficiently accurately the oscillation frequencies of all quasi-breather particles calculated over a certain time interval near $t = t_k$, then they will not be strictly identical. In light of this, we find the root-mean-square deviations $\eta(t_k)$ of the oscillation frequency of various breather particles from the average breather frequency $\bar{\omega}$:

$$\bar{\omega}(t_k) = \frac{1}{N}\sum_{i=1}^{N} \omega_i(t_k), \tag{32}$$

$$\eta(t_k) = \sqrt{\frac{\sum_{i=1}^{N}(\omega_i(t_k)-\bar{\omega}(t_k))^2}{N(N-1)}}.$$

(33)

The larger the value $\eta(t_k)$, the more the quasi-breather solution differs from the exact breather solution, for which $\eta(t_k) = 0$ at any time t_k. Fig. 63 shows the dependence of the root-mean-square deviation η of the quasi-breather on the time of its existence t_k.

The root-mean-square deviation characterizes the measure of data dispersion. In our case, this is the deviation of the frequencies of the peripheral atoms of the model quasi-breather from the frequency of the core of the quasi-breather. Fig. 63 shows that the root-mean-square deviation of the quasi-breather varies from 0.05351804 to 0.07872487, which corresponds to a slight scattering of the frequency of peripheral atoms from the frequency of the core of the model quasi-breather.

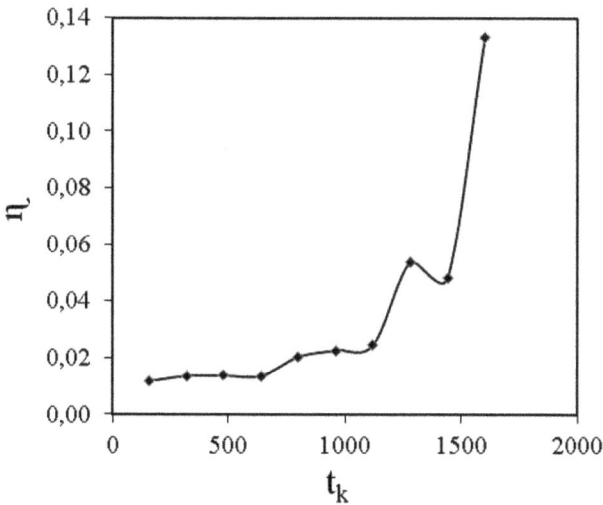

Figure 63: *Dependence of the root-mean-square deviation η of the model quasi-breather on the time of its existence t_k (in ps).*

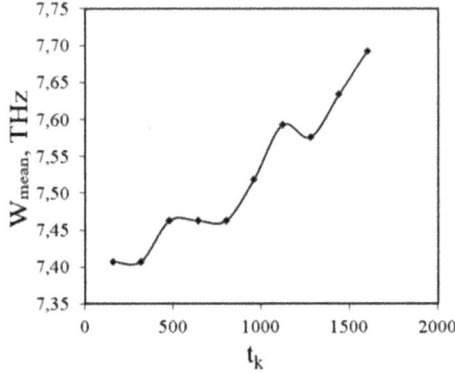

Figure 64: *Dependence of the mean frequency ω_{mean} (in THz) of a model quasi-breather on the time of its existence t_k (in ps).*

The root-mean-square deviation gives an absolute estimate of the spread measure. Therefore, to understand how large the spread is relative to the values themselves (i.e., regardless of their scale), a relative indicator is required. This indicator is called the coefficient of variation and is calculated by the formula:

$$V = \frac{\eta}{\omega_{mean.}}. \tag{34}$$

Table 3. *The exponent of the coefficient of variation V on the lifetime of the quasi-breather t_k (in ps).*

t_k	V
160	0.001575494
320	0.001833693
480	0.001833921
640	0.001821728
800	0.002726615
960	0.003016024
1120	0.003244562
1280	0.007104636
1440	0.006320921
1600	0.01730076

According to this indicator, it is possible to compare the homogeneity of a wide variety of phenomena, regardless of their scale and units of measurement. The table shows the indicators of the coefficient of variation V on the lifetime of the quasi-breather t_k.

Fig. 64 shows the dependence of the average frequency ω_{cp} of the model quasi-breather on the time of its existence t_k.

Figs. 63, 64, and Table 3 show the deviation of the frequency of the peripheral atoms of the quasi-breather from the frequency of the core of the quasi-breather is insignificant. Moreover, the average frequency varies from 7.68688603 THz to 7.83960979 THz, which corresponds to the gap in the phonon spectrum of the Pt$_3$Al crystal (see Fig. 3.49b). Therefore, within the framework of this model of a Pt$_3$Al crystal, one can speak about the closeness of the model quasi-breather to the corresponding exact breather.

This indicates the stability of the resulting discrete breather in model cells and the possibility of excitation in real alloys of the considered composition.

4.2 Statistical analysis of the stability of discrete breathers in monatomic crystals

The statistical approach makes it possible to identify the causes of the destruction of such objects as discrete breathers and to identify the features of these processes. Let us consider discrete breathers with a rigid type of nonlinearity from the standpoint of the concept of quasi-breathers. As objects of study, metals with an fcc structure were chosen: Pt, Ni, Cu, Pd, and Au. This choice of crystals makes it possible to establish a correlation between the DB lifetime and the properties of the materials under consideration. The simulation was performed using the LAMMPS molecular dynamics package. The potentials obtained by the embedded atom method were used as interatomic potentials.

The calculated blocks of crystals, which are cubic cells, contained from $5 \cdot 10^4$ to $5 \cdot 10^5$ atoms. Such sizes of computational cells are sufficient to avoid the influence of boundary conditions on the lifetime of quasi-breathers, taking into account their high degree of spatial localization. Periodic boundary conditions were imposed along three coordinate axes, which made it possible to eliminate the effect of the surface on the dynamics of discrete breathers. The equations of motion of atoms were integrated using a numerical scheme of the fourth order of accuracy with an integration step of 0.5 fs. The main factor determining the DB lifetime in real crystals is the proximity of its frequency to the frequencies of the phonon spectrum. Therefore, the dispersion curves and densities of phonon states were calculated for the crystals under consideration. DBs of the hard type of nonlinearity are localized predominantly in one close-packed atomic row. The X-axis is chosen along the given row. The atoms in the DB core vibrate along a close-packed row in antiphase with their nearest neighbors. The initial conditions for the excitation of an immobile DB with a rigid type of nonlinearity were set according to the ansatz in the works of Dmitriev S.V. and co-authors on the excitation of discrete breathers in monoatomic crystals.

The initial conditions for the excitation of a stationary discrete breather with a rigid type of nonlinearity were set using the S.V. Dmitriev ansatz:

$$x_n^0 = T_n + S_n \,, \; \dot{x}_n^0 = 0, \; y_n^0 = 0, \; \dot{y}_n^0 = 0, \qquad (35)$$

where x_n^0, y_n^0 and $\dot{x}_n^0$, $\dot{y}_n^0$ are the components of the vectors of initial displacements and initial velocities of the n-th atom of the close-packed row of the crystal. All other crystal atoms had zero initial displacements and initial velocities. The functions T_n and S_n describe the vibration amplitudes and displacements of the vibration centers of atoms, respectively. That is, $T_n = (x_{n,max} - x_{n,min})/2$, $S_n = (x_{n,max} + x_{n,min})/2$, where $x_{n,max}$ and $x_{n,min}$ are the maximum and minimum value of the function $x_n(t)$ describing the motion of the n-th atom. These functions looked as follows:

$$T_n = \frac{(-1)^n A}{\cosh[\beta(n-x_0)]}, \quad S_n = \frac{-B(n-x_0)}{\cosh[\gamma(n-x_0)]},$$ (36)

where parameter A determines the DB amplitude, parameter B determines the displacement amplitude of the atomic vibration centers, the parameters β and γ specify the degree of spatial localization of the DB, and x_0 is its initial position. For $x_0 = 0$, we have a DB centered on the atom, and for $x_0 = 1/2$, in the middle between two neighboring atoms.

Fig. 65 shows the initial displacements of atoms from the equilibrium position. An important characteristic of a crystal in the study of discrete breathers is the density of the phonon states of a given crystal. The phonon density of states of crystals was calculated using the molecular dynamics simulation software package (Fig. 17), and it was found that the DB frequency lies outside the phonon spectrum of the crystal. Let's consider the obtained results in more detail. The phonon spectrum of a monoatomic crystal does not have a gap, therefore, the excitation of breathers is possible only if its frequency is split off from the upper boundary of the spectrum. For all metals, the DB frequencies must be greater than a certain value, for example, in a Pd crystal, the DB frequency must be more than 5.6 THz.

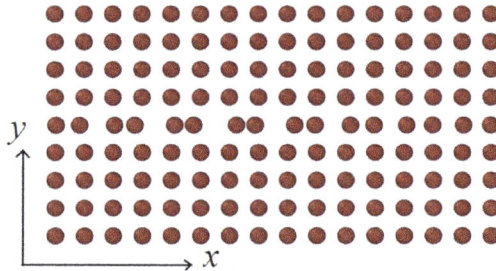

Figure 65: Initial deviations of atoms in an fcc crystal for excitation of a stationary discrete breather with a hard type of nonlinearity, the x-axis is directed along the close-packed <110> crystallographic direction.

For all metals, the phonon densities of states and dispersion curves are calculated (Fig. 66). Further, for each selected metal, such an important characteristic as the lifetime of DBs (Fig. 67) and the dependence of the frequency of vibrations of DB atoms on amplitude (Fig. 68) was studied. By varying the parameter A in equation (36), the value of the amplitude was fixed, and the oscillation frequency of the atoms included in the discrete breather was measured. The resulting dependence is close to linear and corresponds to the hard type of nonlinearity.

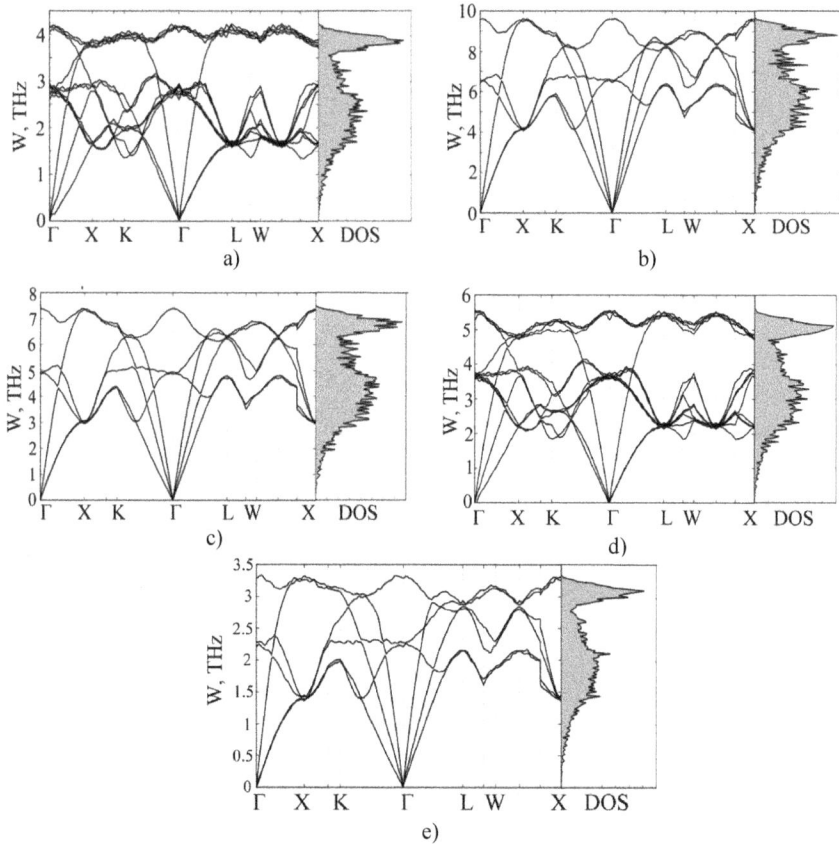

Figure 66: *Dispersion curves and density of phonon states for monoatomic crystals (a) Pt, (b) Ni, (c) Cu, (d) Pd, (e) Au.*

Since it is quite difficult to set ideal initial conditions for the excitation of discrete breathers in realistic models, energy is dissipated into the crystal at the initial stages of the simulation. Consequently, the initial conditions play a decisive role in the lifetime of discrete breathers in crystals.

In the course of the experiments performed, the dependences of the DB lifetime in the considered metals on the initial amplitudes were obtained. Fig. 69 shows examples of such dependencies for Pt, Pd, and Au.

The obtained dependence of the DB lifetime on the amplitude indicates its sufficient stability to variations in the initial atomic amplitudes. At small initial amplitudes, the lifetime of discrete breathers decreases because their frequencies fall into the phonon spectrum of the crystal. At large amplitudes, a significant dissipation of the initial energy transferred to the atoms occurs, which leads to vibrations of neighboring particles and accelerates the dissipation of the DB energy.

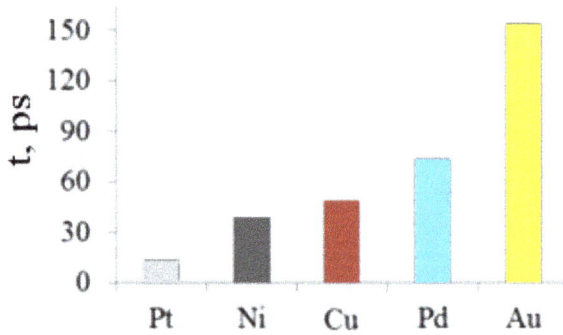

Figure 67: *Lifetime of a discrete breather with a hard type of nonlinearity in the considered fcc metals.*

A study was made on the influence of the initial conditions on its possibility of existence in a crystal.

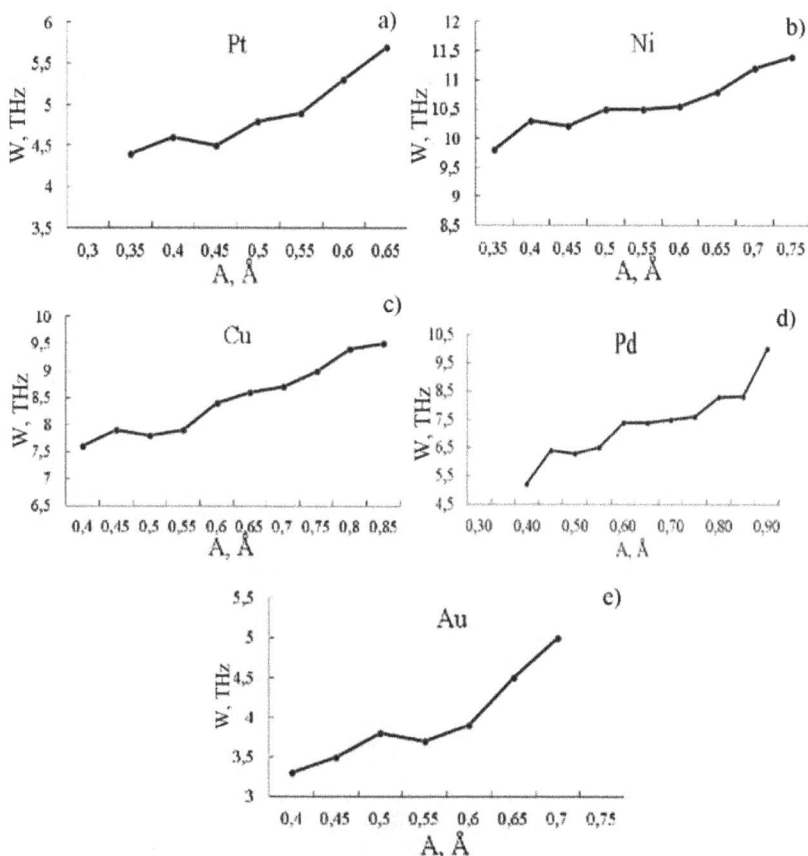

Figure 68: *Dependence of the frequencies of discrete breathers on the amplitudes of atomic vibrations for the metals under consideration.*

Next, we consider the effect of variations in the parameters β and γ on the lifetime of DBs in Pt, Pd, and Au (Figs. 69 and 70).

The parameters β and γ. which determine the degree of DB localization, sufficiently in a wide range of their values provide the conditions for the existence of DBs. Note that the ansatz takes into account the fact that the DB is an object exponentially localized in space, which is ensured by the use of hyperbolic functions in (36). Variation of the ansatz parameters can lead to initial conditions when the energy is rapidly dissipated and the DB is destroyed.

It is obvious that the DB lifetime depends not only on the initial conditions, but also a certain correlation with the properties of crystals is observed. In particular, the dependence of the lifetime on the shear modulus is most pronounced. On the whole, a trend is observed for the crystals under consideration, when the lifetime of the breather increases with a decrease in the shear modulus.

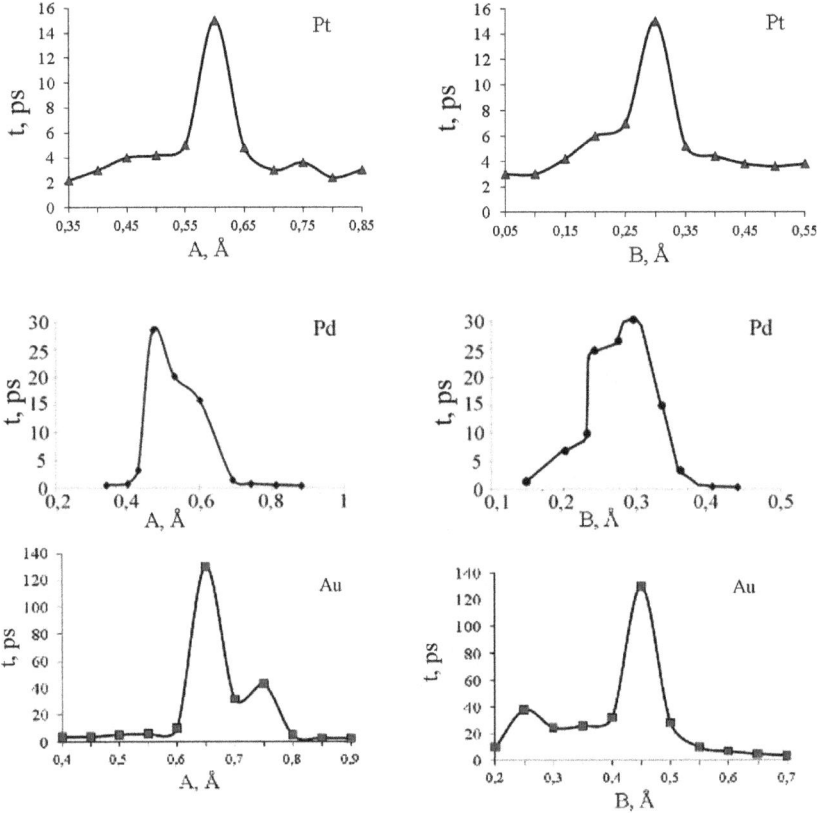

Figure 69: *Dependence of the time of existence of a discrete breather in metals on the value of the initial amplitude of atomic vibrations (parameters A and B of ansatz 36).*

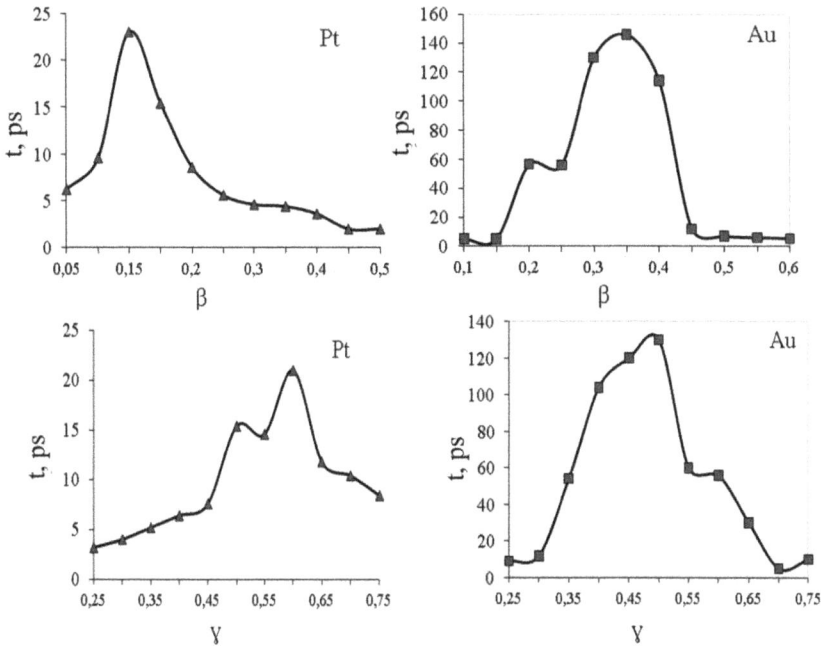

Figure 70: Dependence of the lifetime of a discrete breather on the values of the parameters β and γ.

As already noted, unlike exact discrete breathers, quasi-breathers are not strictly time-periodic dynamic objects and have a finite lifetime. For the metals considered in this paper, the maximum lifetime of quasi-breathers is measured in tens and hundreds of oscillation periods. Note that high-amplitude thermal fluctuations, which are not quasi-breathers, decay after two or three oscillation periods. Thus, quasi-breathers have a lifetime that is orders of magnitude longer than the lifetime of thermal fluctuations, which allows us to consider them as independent physical objects that make a certain contribution to the formation of the properties of the systems in question. The maximum oscillation amplitude of quasi-breathers reaches significant values, in our case, about 0.4 − 0.5 Å.

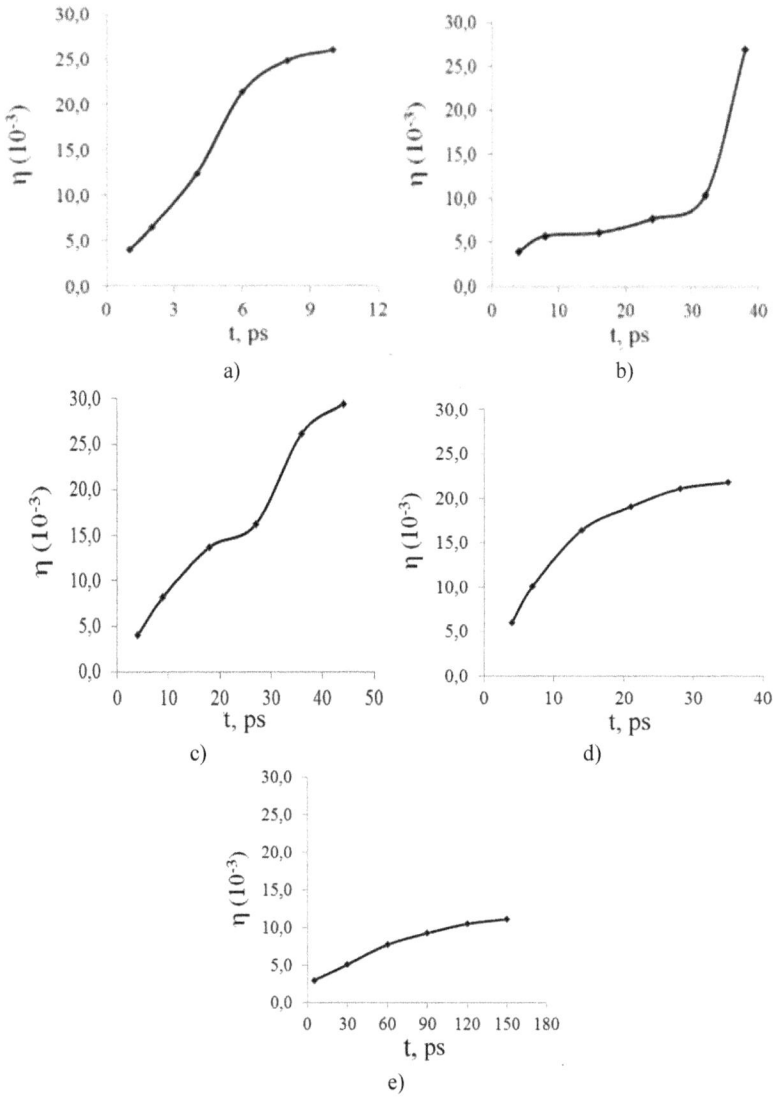

Figure 71: *Dependences of the standard deviation of the quasi-breather η on the lifetime for metals: (a) Pt, (b) Ni, (c) Cu, (d) Pd, (e) Au.*

To characterize the degree of proximity of the quasi-breather to the discrete breather, for all atoms in the core of the quasi-breather, the oscillation frequencies were calculated over a certain time interval near $t = t_k$. As expected, they were not exactly the same. In light of this, we find the root-mean-square deviations $\eta(t_k)$ of the oscillation frequency of various atoms from the average frequency of the breather ω_{cp}.

The larger the value of $\eta(t_k)$, the more the quasi-breather differs from the exact DB, for which $\eta(t_k) = 0$ at any time t_k. Next, we considered a tuple of eight particles that make up the core of the breather.

The dependencies obtained (Fig. 71) show that the faster the standard deviation increases in the studied group of atoms, the faster the breather is destroyed. Not only the rate of rise that is important but also the absolute value.

Fig. 23 shows the dynamics of changes in the average frequency of the quasi-breather over time. For all metals, there is a linear change in frequency with time.

The destruction of the breather occurs now when the difference between the fundamental frequency of the quasi-breather and the upper limit of the phonon spectrum becomes less than the standard deviation. At this moment, the breather begins to actively dissipate energy into the phonon subsystem of the crystal and is destroyed during several periods of oscillations.

Next, the coefficient of variation for quasi-breathers in all the considered crystals was calculated (Table 4). For those crystals in which the lifetime of the quasi-breather was shorter, this index was the largest.

Table 4. *The coefficient of variation on the lifetime of the quasi-breather, for the considered crystals.*

Pt		Ni		Cu		Pd		Au	
t_k	V (10^3)	t_k	V (10^3)	t_k	V (10^3)	t_k	V (10^3)	t_k	V (10^3)
2	1.45	8	0.55	9	1.00	7	1.73	30	1.43
4	2.84	16	0.59	18	1.74	14	2.88	60	2.19
6	5.07	24	0.76	27	2.11	21	3.40	90	2.65
8	6.04	32	1.04	36	3.53	28	3.83	120	3.05
10	6.44	38	2.73	44	4.02	35	4.03	150	3.28

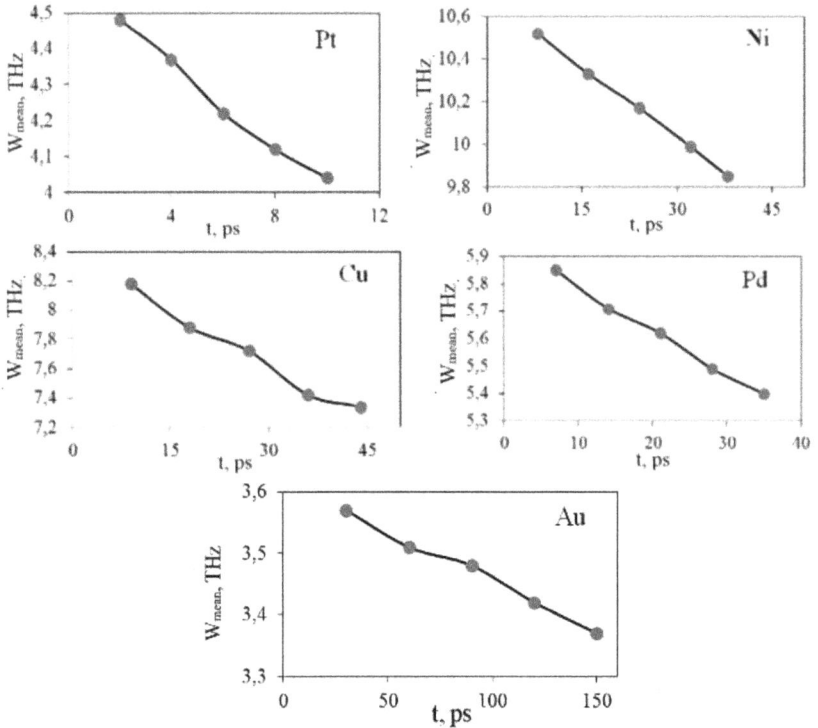

Figure 72: *Dependence of the average frequency ω_{cp} of the quasi-breather on the lifetime for metals: Pt, Ni, Cu, Pd, and Au.*

The data obtained in this section expand our understanding of the properties of quasi-breathers in pure fcc metals, they can be useful in setting up experiments on the indirect observation of quasi-breathers in real crystals, as well as in explaining the effects of energy dissipation under intense external influences on metal materials.

4.3 Statistical characteristics of a quasi-breather with a hard type of nonlinearity in a CuAu crystal and monoatomic Cu and Au

In this section, the statistical characteristics of quasi-breathers with a hard type of nonlinearity are calculated in a model CuAu crystal, as well as in Cu and Au single crystals.

The model under consideration was a bulk crystal containing 48,000 atoms (see Figure 73). Periodic boundary conditions were imposed along all directions.

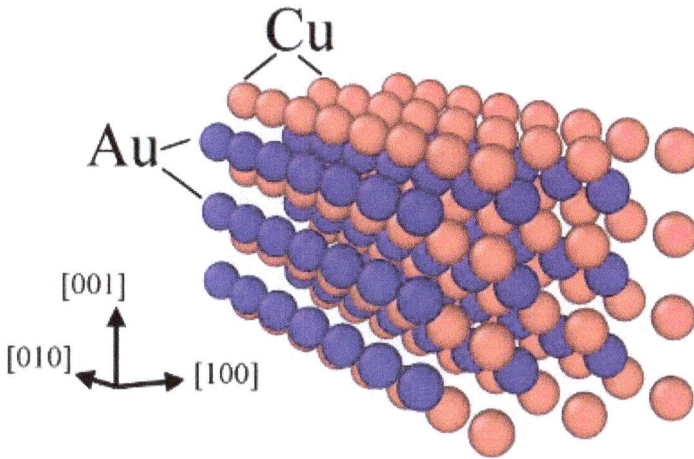

Figure 73: Fragment of a CuAu crystal.

Unlike exact discrete breathers, quasi-breathers are not strictly time-periodic dynamic objects, although they are localized in space. In light of this, we will find the root-mean-square deviations $\eta(t_k)$ of the oscillation frequency of various breather particles from the average breather frequency ω_{cp}.

Fig. 74 shows the dependence of the root-mean-square deviation η of a quasi-breather on its lifetime t_k.

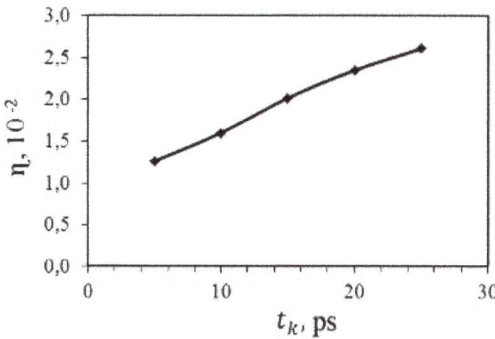

Figure 74: Dependence of the root-mean-square deviation η of a quasi-breather the lifetime t_k.

The standard deviation characterizes the measure of data dispersion. In our case, this is the deviation of the frequencies of the peripheral atoms of the model quasi-breather from the frequency of the core of the quasi-breather. Figure 75 shows that the root-mean-square deviation of the quasi-breather varies from 0.01261065 to 0.02610272, which corresponds to an insignificant scattering of the frequency of peripheral atoms from the frequency of the core of the model quasi-breather.

Fig. 75 shows the dependence of the average frequency ω_{cp} of the quasi-breather on the time of its existence t_k.

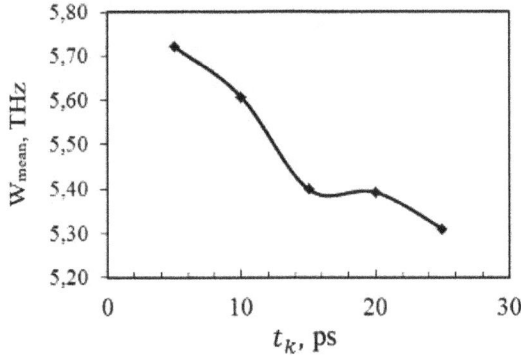

Figure 75: *Dependence of the average frequency ω_{cp} of a model quasi-breather on the time of its existence t_k.*

The standard deviation gives an absolute estimate of the spread measure. Therefore, to understand how large the spread is relative to the average values, a relative indicator was calculated, - the coefficient of variation. According to this indicator, it is possible to compare the homogeneity of a wide variety of phenomena, regardless of their scale and units of measurement. Table 5 shows the indicators of the coefficient of variation V from the lifetime of the quasi-breather t_k.

Table. 5. *Dependence of the coefficient of variation V on the lifetime of the quasi-breather t_k CuAu.*

t_k	V
5	0.00220377958477173
10	0.00286022261352874
15	0.00374032727309376
20	0.00435946841613917
25	0.00491586066089457

The lifetime of the considered quasi-breathers was divided into five equal parts. Thus, five points were obtained for the analysis of the statistical characteristics of the breathers. That is, there was a sample of five elements - the frequencies of quasi-breathers, see Table 5.

The following are the results (Figure 76) for CuAu compared to the monatomic components of this alloy. And a sample of medium frequencies was made (Table 6).

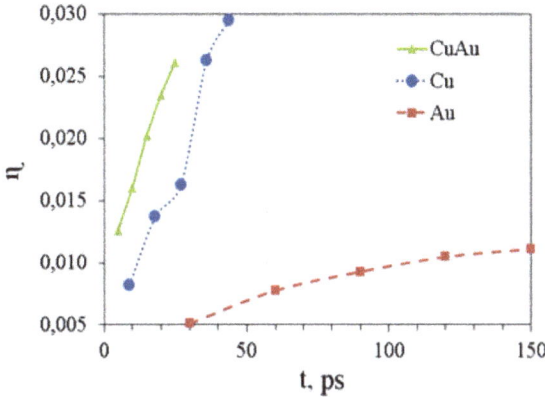

Figure 76: *Dependence of the root-mean-square deviation of vibrational frequencies of quasi-breathers atoms on the time of their existence for CuAu, Cu, Au.*

Table 6. *Quasi-breather frequency sampling.*

Model	ϖ_1, THz	ϖ_2, THz	ϖ_3, THz	ϖ_4, THz	ϖ_5, THz
CuAu	5.72229	5.60811	5.39909	5.39343	5.30990
Cu	8.18301	7.87534	7.72063	7.42053	7.33631
Au	3.56532	3.51414	3.48136	3.42447	3.37433

Next, a statistical series of absolute frequencies was constructed for this sample, i.e. a sequence of pairs of numbers $(\omega_1^*, n_1^*), (\omega_2^*, n_2^*), \ldots, (\omega_m^*, n_m^*)$, where ω_k^* is the center of the kth grouping interval and n_1^* is the number of sample elements that fall into k-th interval. Numbers n_k^* ($k = 1, \ldots, m$) are called absolute frequencies. We find a minimum and maximum sample element, they correspond to the extreme values for each model in Table 1. We find the length of the grouping interval according to the following formula:

$$h = (\omega_{max} - \omega_{min})/m. \tag{37}$$

Find the right boundaries of the grouping intervals:

$$\omega_k = \omega_{min} + kh \ (\kappa = 1, ..., 5). \tag{38}$$

Find the centers $\omega*_k$ of the grouping intervals according to the formula:

$$\omega^*_k = \omega_k - h/2 \ (\kappa = 1, ..., 5). \tag{39}$$

For each grouping interval (ω_{k-1}, ω_k) we find the number n_k* of sample elements that fall into this interval. Each element of the sample must be assigned to one and only one interval, and if the value of the element falls on the border of the interval, then we refer it to the interval with the lowest number. The minimum element is always assigned to the first interval and the maximum to the last. The results obtained are given in Table 7.

Table 7. *Auxiliary table of statistical data.*

Model	Interval number k	Interval center ω_k^*, THz	Interval limits, THz
CuAu	1	5.35114	5.30990...5.39238
	2	5.43361	5.39238...5.47485
	3	5.51609	5.47485...5.55733
	4	5.59857	5.55733...5.63981
	5	5.68105	5.63981...5.72229
Cu	1	7.42098	7.33631...7.50565
	2	7.59032	7.50565...7.67499
	3	7.75966	7.67499...7.84433
	4	7.92900	7.84433...8.01367
	5	8.09834	8.01367...8.18301
Au	1	3.39343	3.37433...3.41253
	2	3.43162	3.41253...3.45072
	3	3.46982	3.45072...3.48892
	4	3.50802	3.48892...3.52712
	5	3.54622	3.52712...3.56532

We also build a grouped statistical series of relative frequencies, which is a sequence of pairs of numbers $(\omega_1*, n_1*/n), (\omega_2*, n_2*/n), ..., (\omega_m*, n_m*/n)$, where n_k*/n - relative frequencies and n is the sample size (see Table 5).

Table 8. *Grouped statistical series of relative frequencies.*

CuAu	ω_k^*, THz	5.35114	5.43361	5.51609	5.59857	5.68105
	n_k^*/n	0.20000	0.40000	0.00000	0.20000	0.20000
Cu	ω_k^*, ТГц	7.42098	7.59032	7.75966	7.92900	8.09834
	n_k^*/n	0.40000	0.00000	0.20000	0.20000	0.20000
Au	ω_k^*, THz	3.39343	3.43162	3.46982	3.50802	3.54622
	n_k^*/n	0.20000	0.20000	0.20000	0.20000	0.20000

Based on Table 8, we will construct polygons of relative frequencies for each of the crystal models (see Fig. 77).

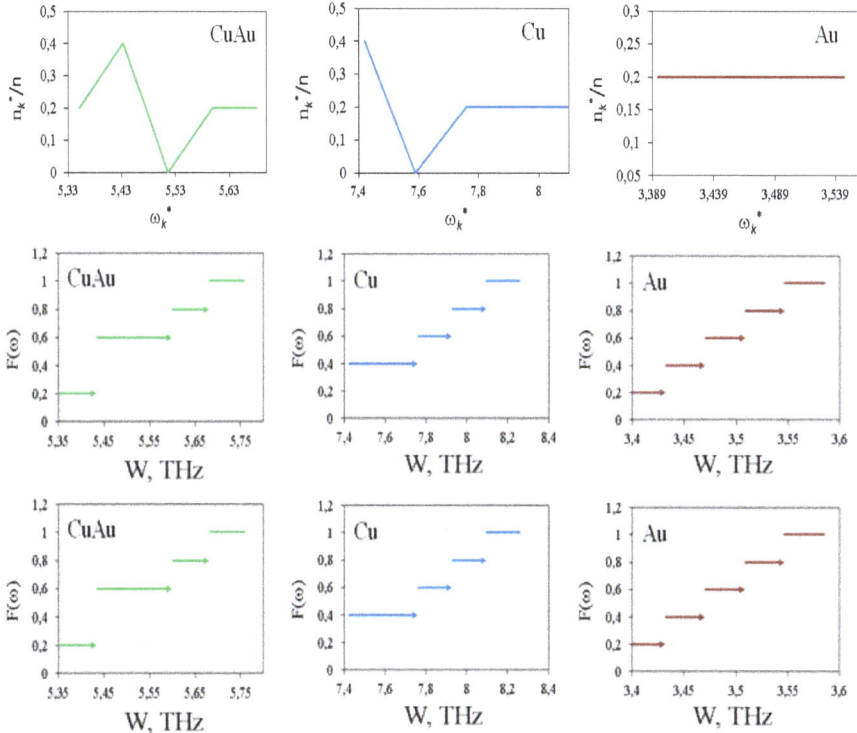

Figure 77: *Polygons of relative frequencies of discrete breathers in CuAu, Cu, Au, ω_k^* in THz.*

To complete the statistical picture of the characteristics of quasi-breathers, we will evaluate the mathematical expectation and variance, and also construct empirical distribution functions.

The estimate of the mathematical expectation (sample mean) of an ungrouped sample is calculated by the formula:

$$M^* = \frac{1}{n}\sum_{k=1}^{n} \omega_k. \tag{40}$$

Estimation of the variance of an ungrouped sample is carried out according to the formula:

$$D^* = \frac{1}{n-1}\sum_{k=1}^{n}(\omega_n - M^*)^2.$$

For the models we are considering, we obtained the values given in Table 9.

Table 9. *Mathematical expectation and dispersion for model crystals.*

Model	M^*	D^*
CuAu	5.486563	0.029469
Cu	7.707168	0.118647
Au	3.471924	0.005593

For clarity, we construct empirical distribution functions $F(\omega)$ (see Fig. 78).

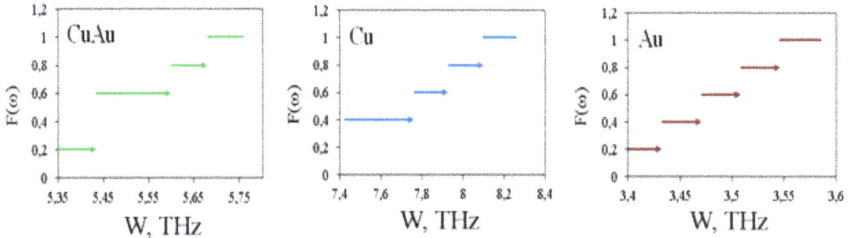

Figure 78: *Empirical distribution functions for CuAu, Cu, Au.*

The obtained statistical data show the process of energy dissipation by breathers over the entire interval of their lifetime. The destruction of quasi-breathers occurs at the moment when the root-mean-square deviation exceeds the difference between the average frequency of the breather and the nearest boundary of the phonon spectrum of the crystal. In this case, this process may not be uniform, which is primarily due to the properties of the crystals, as well as the method of excitation of the breathers. The existence of the quasi-breather was the longest in the Au crystal, while the process of energy dissipation proceeded uniformly throughout the entire period of its existence. In Cu and CuAu crystals, the main part of the energy was dissipated during the initial periods of the breather existence, which is probably due to the initial conditions for the excitation of these objects.

To excite quasi-breathers, two different approaches were used, based on different functions of displacement of atoms from the equilibrium position. All main statistical characteristics of quasi-breather frequencies have been calculated: standard deviation of atomic frequencies, average frequencies of a quasi-breather, polygons of relative frequencies, mathematical expectation, variance, and empirical distribution functions. It has been established that the root-mean-square deviation of the oscillation frequencies of quasi-breathers atoms, that is, the degree of their quasi-breather property, increases with time, and the average frequency of their oscillations decreases, approaching the upper limit of the phonon spectrum. Quasi-breathers are destroyed when the standard deviation of the oscillation frequencies exceeds the difference between the average frequency of the breather and the nearest boundary of the phonon spectrum of the crystal. The obtained statistical data make it possible to describe the process of DB degradation over time. The described approaches must make it possible to establish that quasi-breathers, which have a shorter lifetime, dissipate energy at the initial stages of the observation period, which is determined both by the breather excitation method and by the properties of model crystals.

Chapter 5. Solitary waves in FCC crystals

5.1 Supra-transmission effect in biatomic crystals of stoichiometry A_3B

Methods for modifying surface layers of materials are very often based on surface treatment with high-intensity external influences in the form of a plasma discharge, annealing, current pulses, etc. Energy flows from the surface of crystals affect the structural and energy transformations of materials, thus providing a modification of the near-surface layers of matter.

In this section of the work, we consider the effect of energy transfer upon periodic exposure of the crystal surface to the stoichiometric composition A_3B, which has a band gap in the phonon spectrum of the crystal. The impact was carried out in a wide range of frequencies both included in the phonon spectrum and outside the phonon spectrum of the crystal. The effect of energy transfer at frequencies outside the phonon spectrum of the crystal is called nonlinear supra-transmission. Interest in this effect does not fade both for relatively simple nonlinear systems and more complex systems and materials. In the classical approach to supra-transmission, there is an initial value of the amplitude at which this effect occurs. However, in the example of deformed graphene, the possibility of energy transport through nonlinear supra-transmission is shown without restrictions on the minimum value of the impact amplitude. This motivates the study of this effect for various crystals and the search for energy transfer mechanisms in nonlinear systems.

This mechanism is interpreted by the excitation of nonlinear localized modes of large amplitude near the impact zone - discrete breathers. The paper considers a model of a Pt_3Al crystal. Interest in this alloy is due to the prospect of its use in the composition of superalloys, as well as its resistance to high temperatures. In addition, we have shown above the possibility of excitation of discrete breathers in a given material by a particle flow even at thermodynamic equilibrium.

Periodic boundary conditions were imposed on the fcc crystal model (Fig. 79) along the X and Y axes, and free ones along the Z axis. The resulting model was divided into three blocks. Block I consisted of 3-4 layers of atoms, which carried out vibrations according to a harmonic law in accordance with the ranges of frequencies and amplitudes of atomic vibrations. Periodic action was applied to all atoms from block I. Next was block II - an energy absorber, according to which the absorbed energy by the crystal was estimated. The effect of the size of zone II on the amount of absorbed energy has not been evaluated. The study aimed to determine the dependence of the total energy E of the atoms of zone II on the frequency ω of the periodic action and the type of vibrations. In part III of the computational cell, a block of 4-5 layers of atoms, rigidly fixed, acting as a damper, was distinguished. This ensured that the entire model of the Pt_3Al crystal did not move, and made the model closer to the real crystal. Fig. 1b shows the density of phonon states of the crystal model under consideration.

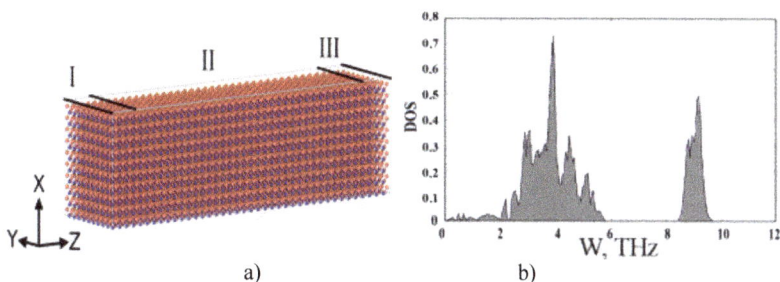

Figure 79: (a) View of a three-dimensional model of a Pt_3Al crystal, the X axis is directed along the crystallographic direction <100>, Y - <010>, Z - <001>. The number I indicates the area of periodic impact, II - the area of energy absorption, III - rigidly fixed atoms; (b) the density of phonon states of the Pt_3Al crystal.

Periodic exposure was carried out according to the following laws:

$$Z_1(t)=A\sin(\omega t) \tag{42}$$

$$Z_2(t)=A(\sin(\omega t))^2 \tag{43}$$

$$Z_3(t)=A|\sin(\omega t)| \tag{44}$$

where for all cases A is the amplitude of the external influence, and ω is the oscillation frequency of region I in Fig. 79a. These harmonic laws were entered through commands built into LAMMPS mathematical functions and variables. Oscillations were made along the Z axis with frequencies from 0.2 to 15 THz, and with different amplitudes from 0.05 to 0.5 Å. Such a range makes it possible to cover the entire spectrum of low-amplitude atomic vibrations for the crystal under consideration.

When performing calculations using the LAMMPS package, the following parameters were used:
- the integration step was 0.001 ps, while the estimated time of each launch was 1 ps;
- design temperature was 970 K;
- temperature control scheme used.

During the calculations for each of the above periodic laws, 52 runs were performed corresponding to the frequency range from 0.2 to 15 THz for each amplitude from 0.05 to 0.5 Å. Thus, the calculation time took from four to five hours for each amplitude value. At the same time, repeated launches were made to compare the results obtained; they showed sufficient reliability of the calculations.

The absorbed energy for zone II atoms was calculated as follows: using the *compute* command, the kinetic energy of atoms from block II was calculated, then the resulting value was divided by the number of atoms in this block, and then the results were averaged. Such calculations were carried out for each frequency value.

The absorbed energy was recorded depending on the frequency of exposure and amplitude. The results obtained for the action according to the law (42) are shown in Fig. 80. Based on this graph, it can be noted that for amplitudes less than 0.2 Å, the effect of energy transfer to the crystal was absent in the band gap of the phonon spectrum (Fig. 79b). For amplitudes of 0.2 angstroms or more, energy is transferred to the crystal, including at frequencies in the band gap of the phonon spectrum. As the amplitude increases, the absorption peak shifts deeper into the forbidden frequencies of the phonon spectrum of the crystal, which indicates an increase in the fraction of nonlinear modes in the process of energy transfer. The active excitation of breathers occurs close to the area of action; we obtained similar results earlier for two-dimensional models of this alloy. Note that there is a threshold value of the impact amplitude, from which this effect begins to manifest itself, which is typical for the classical interpretation of the supra-transmission effect. Also characteristic is the peak at exposure frequencies of 6 THz. The nature of the absorption burst at a given frequency will be discussed below, taking into account the analysis of the results of the action using equations (43) and (44).

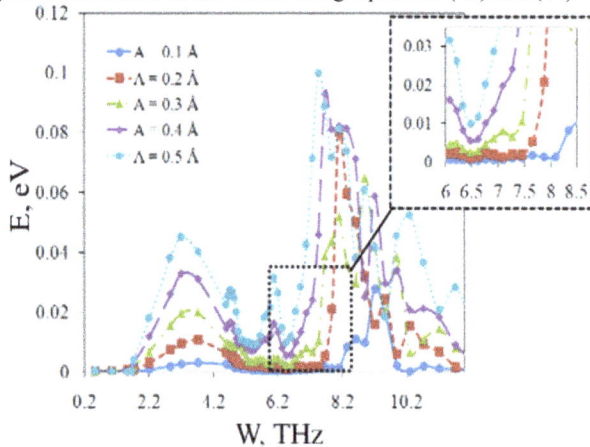

Figure 80: *Dependence of the energy absorbed by the computational cell per atom per picosecond on the frequency of the external action and the amplitude according to the law (42), the scale is increased for frequencies from 6 to 9 THz.*

Figs. 81 and 82 show the results of actions according to harmonic laws (43) and (44). The graphs show that the amount of energy transferred to the crystal is somewhat less than when exposed to the formula (42). The reason for the discrepancies in the results may be that the excitation of quasi-breathers near the area of influence occurs more slowly in these cases because the form of action differs more strongly from the excitation parameters of quasi-breathers.

Since the results obtained for equations (42) - (44) have some differences, then the behavior of the first layer of atoms near the impact area is analyzed further. To do this, one Pt and Al atom were selected in this layer and their dynamics were monitored over time for the entire frequency range. For example, Fig. 83 shows the change in the Z coordinate of atoms relative to the equilibrium position for a frequency of 6 THz and an amplitude of 0.4 (Fig. 83, a, b), for a frequency of 8 THz and an amplitude of 0.2 Å (Figure 83, c, d).

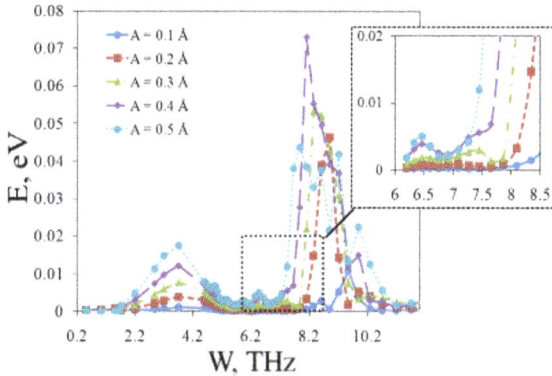

Figure 81: Dependence of the energy absorbed by the computational cell per atom per picosecond on the frequency of the external action and the amplitude according to the law (43), the scale is increased for frequencies from 6 to 9 THz.

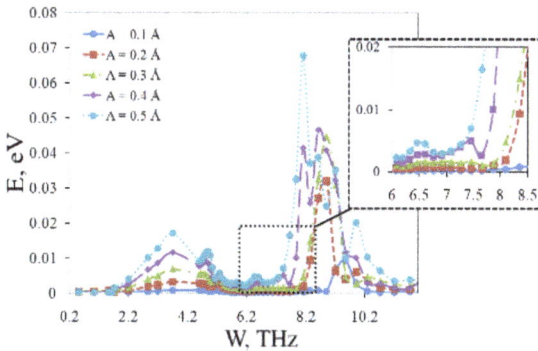

Figure 82: Dependence of the energy absorbed by the computational cell per atom per picosecond on the frequency of the external action and the amplitude according to the law (44), the scale is increased for frequencies from 6 to 9 THz.

For frequencies close to the optical branch of the FS, according to the harmonic law (42), oscillations on the Al atom are more actively excited, which indicates the excitation of discrete breathers, while for expressions (43) and (44), such activity is not observed. In this case, a local expansion occurs near the area of influence, which indicates the excitation of discrete breathers. Considering the energy absorption peak at a frequency of 6 THz, we note that it is due to the active involvement of heavy sublattice atoms in vibrations since the impact frequency is close to the upper limit of the acoustic part of the crystal. Thus, it follows from Fig. 5b that Pt atoms, when formula (42) is employed, perform high-amplitude vibrations. However, as already mentioned above, such large values of the external action amplitudes lead to the rapid destruction of the near-surface layer of the model crystal, which happened after 1.5 ps impact. Similar results were obtained for the entire spectrum of frequencies and amplitudes of external periodic action according to equations (42) - (44).

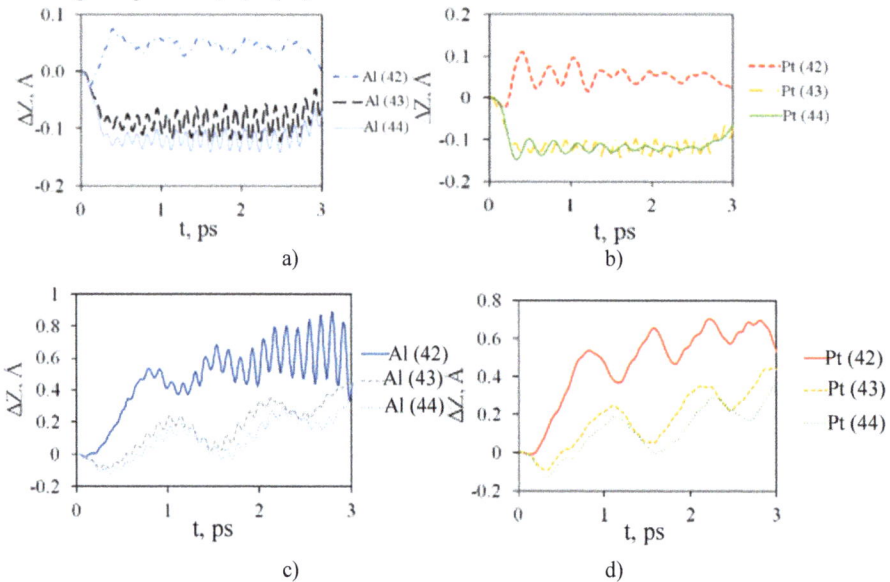

Figure 83: *Change in the Z coordinate of Pt and Al atoms relative to the equilibrium position at the boundary of the impact area, when exposed to harmonic laws (42), (43), and (44); (a) and (b) impact frequency 6 THz, amplitude 0.4 for the corresponding impact, (c) and (d) impact frequency 8 THz, amplitude 0.2 for the corresponding impact.*

In this section, the mechanism of nonlinear supratransmission for various forms of external action is considered by the method of molecular dynamics for a Pt₃Al crystal. It was shown that energy transport through this mechanism is possible along the directions corresponding to the crystallographic directions of the existence of a quasi-breather in a crystal. The results obtained indicate that the contribution of quasi-breathers to the energy transfer through the crystal increases with an increase in the amplitude of the action. The results of the study can be useful in the laser processing of materials and surface treatment with low-energy plasma, as well as in radiation materials science.

5.2 Solitary waves in biatomic crystals of stoichiometry A₃B

In this part of the work, the excitation of solitary waves in crystals of the stoichiometric composition A_3B is considered by computer simulation and the role of discrete breathers in this process is established. Ni_3Al and Pt_3Al were chosen as working crystals. In the Pt_3Al crystal, there is a band gap in the phonon spectrum, which provides the possibility of the existence of discrete breathers with a soft type of nonlinearity in it, while in Ni_3Al there is no such gap, and, accordingly, there are no conditions for the excitation of discrete breathers.

The model we are considering is a bulk fcc crystal of A_3B stoichiometry containing up to $2.5 \cdot 10^6$ particles interacting through the potential obtained by the embedded atom method (EAM-potential) for Pt_3Al and Ni_3Al. The simulation was carried out using the LAMMPS package. The cell size was varied over a wide range in order to study the mechanisms of excitation, propagation, and interaction of solitary waves. In the model, periodic boundary conditions were imposed along the X and Y axes, and free ones along the Z axis. Note that the cell size was changed only along the Z axis, thereby achieving linear dimensions up to 2.4 micrometers. Along the X and Y axes, the cell sizes were 3.5 nm for Ni_3Al and 4 nm for Pt_3Al.

The simulation process consisted of the preparatory stage of modeling and splitting the cell into blocks (Fig. 79a), as in the previous paragraph. Periodic exposure was carried out according to the harmonic law $z(t) = A \sin(wt)$ along the Z axis with frequencies from 0.2 to 20 THz, as well as with different amplitudes from 0.05 to 0.5 Å. Such a range makes it possible to cover the entire spectrum of low-amplitude atomic vibrations for the crystals under consideration. The exposure time was carried out in the range of 1.5 - 8 ps.

For the considered models of Ni_3Al and Pt_3Al crystals, the density of phonon states was calculated (Fig. 84). The LAMMPS software package was used in the calculations, which includes the procedures necessary for these purposes, based on the Fourier transform of the autocorrelation functions of atomic displacements versus time. This makes it possible to measure the frequencies of the external action and the eigenfrequencies of the phonons of the crystals. It is also important when assessing the contribution of discrete breathers to the process of generating solitary waves.

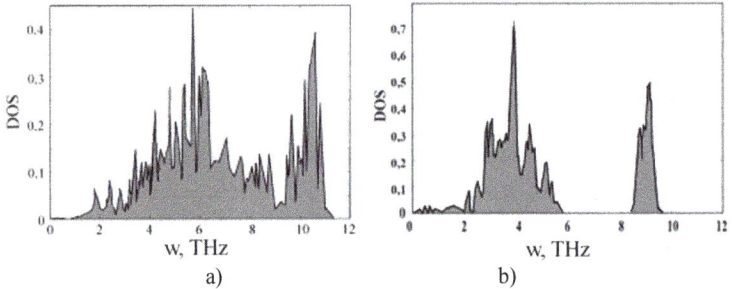

Figure 84: Densities of phonon states of Ni₃Al (a) and Pt₃Al (b) crystals.

As a result of the action in the specified ranges of frequencies and amplitudes, it can be said that there are no solitary waves in the Ni$_3$Al crystal. For example, we present the results of an experiment with the following parameters of external periodic exposure ω = 7.14 THz, A = 0.2 Å, and exposure time t = 2 ps (Fig. 85).

The behavior of the Pt$_3$Al crystal differed from that of Ni$_3$Al, and clearly defined solitary waves formed in it (Fig. 85). The nature of these waves depended on the amplitude, frequency, and duration of the external periodic impact. Since such an effect was not observed in Ni$_3$Al, only the Pt$_3$Al crystal model will be studied next.

Let us consider the mechanism of formation of the obtained solitary waves. In the previous section, we showed that periodic action at frequencies close to the natural frequencies of discrete breathers causes their excitation near the action area, which is expressed in a sharp increase in the amplitude of Al atoms.

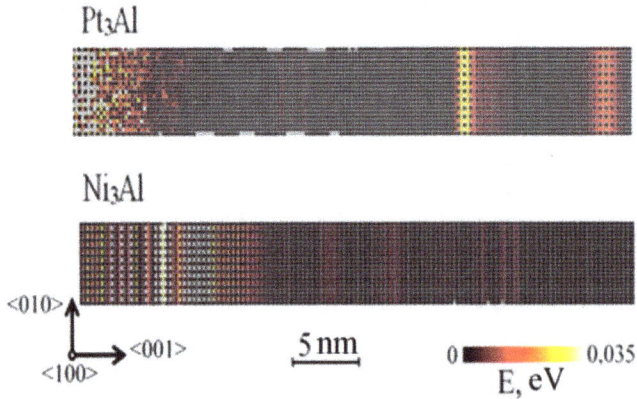

Figure 85: Kinetic energy distribution in Pt₃Al and Ni₃Al crystals 8 ps after the start of the experiment.

Fig. 86 shows the plot of the Z coordinate of the Al atom relative to its equilibrium position in the boundary layer with the impact area for impact amplitudes of 0.1 and 0.2 Å. The change in the coordinate has a wave-like character, at the moment of reaching the maximum value, a solitary wave begins to form, discrete breathers give off their energy, despite the ongoing external influence, then the process repeats. It can be seen in Fig. 86 that two solitary waves were formed for an impact amplitude of 0.2 Å, which are also clearly visible in Figure 85. Also, these results confirm the boundary value of the amplitude for the appearance of the supra-transmission effect. At an amplitude of 0.1 Å, DBs were not formed, and the energy was practically not transferred because the impact frequency was outside the PS of the crystal.

An increase in the amplitude of the external action leads to the rapid destruction of the near-boundary region, however, high-amplitude vibrations of atoms arise in the nearest ordered part of the crystal and excite solitary waves in the same way.

By varying the exposure frequency, it was found that this effect occurs only at frequencies outside the PS of the crystal, including frequencies above the optical branch of the spectrum. In this case, the greater the value of the impact frequency, the greater the amount of energy that can be concentrated on a solitary wave.

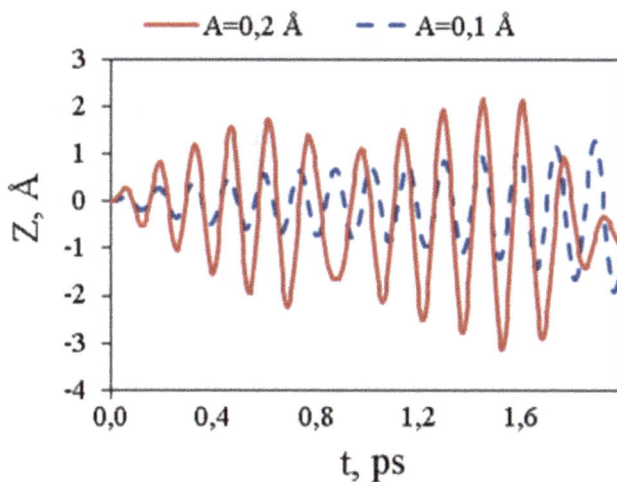

Figure 86: Change in the Z coordinate of the Al atom relative to the equilibrium position at the boundary of the impact area.

Let us consider in more detail the nature of the energy distribution for a single wave excited with the following parameters of the external action: $\omega = 8$ THz, $A=0.2$ Å and exposure time t = 1.5 ps. In this case, only one wave was formed, which had a bell-shaped form (Fig. 87). Note that the main fraction of energy is concentrated on Pt atoms, while Al atoms account for an insignificant part of it. Fig. 87a shows the energy profile of the wave, which makes it possible to estimate the amount of stored energy, as well as a graph of the forces acting on atoms along the direction of wave propagation, Figure 87.

The energy estimate of a concentrated wave for this model (4x4 nm) is about 20-50 eV. Thus, such a wave propagating through a crystal can have a significant effect on the processes occurring in crystals at the atomic level.

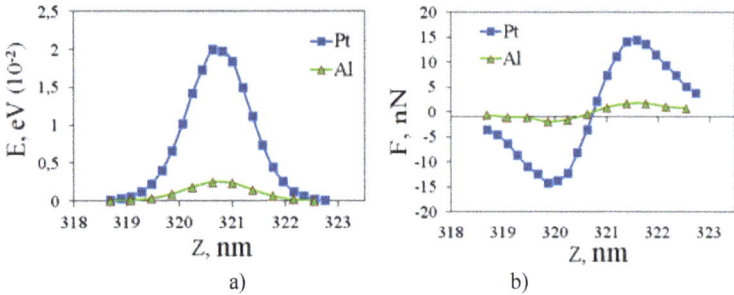

Figure 87: *Distribution of energy (a) and force (b) in the soliton for the alloy components along the cell.*

Next, we consider the well-known theoretical approaches to the description of solitons. In the theory of solitary waves in solids, the solution of the sine-Gordon equation of the form is known:

$$U(q, t) = \frac{2\alpha^2 \cosh \varepsilon^2}{1 + \alpha^2 \cosh \varepsilon^2} \tag{45}$$

where $\varepsilon = q - Vt$, α is the amplitude multiplier, q is the coordinate, V is the speed of the soliton, and t is the time.

Selecting the appropriate values of the parameters of this solution, the profile of a solitary wave is obtained, which is in good agreement with the profile of a soliton in a Pt_3Al crystal (Fig. 88).

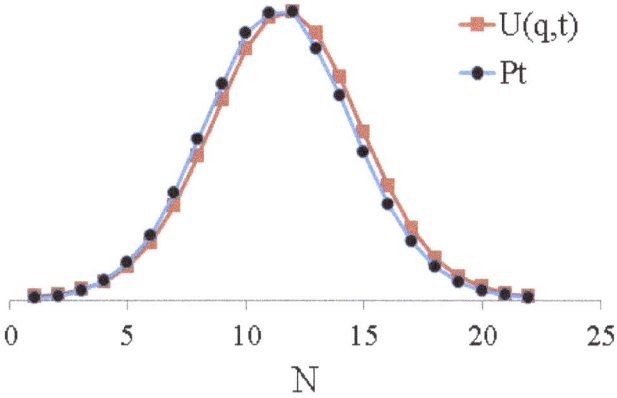

Figure 88: *Soliton profile in Pt₃Al and the profile described by function (45). The amplitude is plotted along the vertical axis, and the number of atoms in the row is plotted along the horizontal axis.*

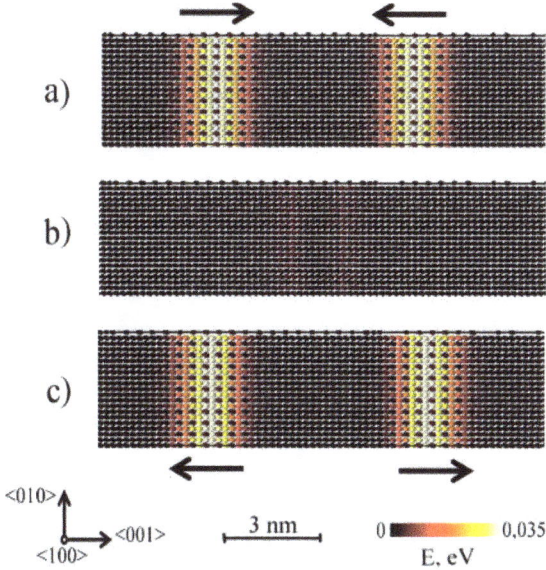

Figure 89: *Collision of two solitary waves: (a) before the collision, (b) moment of the collision, (c) after the collision.*

We can say that the obtained solitary wave belongs to the classical waves of the soliton type. Its profile is bell-shaped, it is not linear and spreads at a constant speed. An estimate of the propagation velocity of such a wave gave a value of the order of $4.5 \cdot 10^3$ m/s.

Fig. 89 shows the distribution of kinetic energy along the crystal when two solitary waves collide. The shape and speed of the waves remained unchanged.

Such waves also turned out to be resistant to heating. The wave, without a significant change in its profile and velocity, can overcome the heated regions of the crystal, at least at temperatures up to 300 K.

It was not possible to establish the maximum possible distance that a soliton can travel within the framework of the models under consideration due to significant computational time costs. The maximum distance traveled by such a wave in the models was more than 11 μm when rigid boundary conditions were imposed on the cell along the Z axis, after the launch of a solitary wave. The soliton was reflected from the rigidly fixed atoms along the edges of the cell and wandered along it without any signs of energy dissipation and change in the speed of movement.

Thus, the obtained solitary waves are capable of propagating thousands of nanometers deep into a defect-free crystal without changing their shape and speed. The total amount of energy carried by the wave is determined by the number of rows of atoms involved in the oscillations, we can talk about tens and hundreds of electric volts. This mechanism of energy transport through the crystal, by means of solitary waves, seems to be one of the most efficient, and the mechanism for generating such waves is relatively simple.

Chapter 6. Post-cascade shock waves in fcc crystals

6.1 Effect of shock waves on structural changes occurring in the depleted zone of a crystal

One of the most important tasks of radiation materials science is to predict the behavior of structural materials under conditions of intense irradiation with high-energy particles to improve their radiation resistance.

It is known that the damaged region formed as a result of the passage of a cascade of atomic collisions has a differentiation in the distribution of atoms. The center of this region is a depleted zone surrounded by a cloud of displaced atoms. Due to the processes of channeling and focusing of atomic collisions, anisotropy may appear in the distribution of interstitial atoms, which prevents their annihilation with vacancies. As a result, vacancy clusters, which are present in excess, begin to rearrange into various clusters - dislocation loops, stacking fault tetrahedra, etc.

Along with the traditional methods of physical experiments, radiation defects are successfully studied by computer simulation methods. These methods allow for obtaining results that cannot be achieved experimentally today.

For example, when studying the stability and formation mechanisms of various vacancy clusters in fcc metals, it was shown that during relaxation, small clusters combine and form stacking fault tetrahedra (SFTs). Nevertheless, despite a large number of different works on this topic, the possibility of propagation of post-cascade shock waves through the crystal is not taken into account during studying the processes of SFT formation in depleted zones. These shock waves are formed due to the mismatch between the times of thermalization of atomic vibrations in a certain finite region and the removal of heat from it. As a result of a sharp expansion of a strongly heated region, an almost spherical shock wave is formed. As our past studies have shown, these waves can have a significant effect on the processes of formation, clustering, and migration of defects.

This section presents the results of studies aimed at revealing the features of structural transformations occurring in the depletion zone belonging to the fcc crystal (by the example of nickel) under the influence of post-cascade shock waves.

During the study, the shock wave was created as follows. A single layer of boundary atoms of the computational cell was singled out, after which these atoms were assigned a velocity equal in magnitude (one and a half times higher than the velocity of longitudinal elastic waves in nickel) and directed along the X-axis. The atomic displacements propagating according to the relay-race principle were a traveling wave, the front width of which is several interatomic distances, and the vibration amplitude significantly exceeds the thermal vibrations of atoms, which is typical for waves generated by the cascade region.

At the initial stage, structural rearrangements occurring in the depleted zone of the simulated crystal were studied at different temperatures. As a result of modeling, it was found that vacancies begin to combine into clusters of various sizes, with the subsequent formation of various stacking fault tetrahedra (see Fig. 90). As a rule, tetrahedra were partially unfinished. The stacking faults formed in the course of the simulation correspond to atoms with a local hcp environment. To estimate the proportion of such atoms, we used structural analysis, which consists in identifying the local environment of particles using the Ackland-Jones method of angles and bonds.

As the concentration of vacancies increases, more significant structural changes are observed in the simulated crystal. Thus, at a 15% concentration of vacancies, local amorphization was observed. With significant heating and subsequent hardening, the presence of excess free volume contributes to the formation of a grain structure (see Fig. 91a). A further increase in the created vacancies leads to pore formation in the crystal (see Fig. 91b).

For the constructed molecular dynamics model, the described changes in the crystal structure are most characteristic. Therefore, it is of interest to study the effect of a shock wave on structural transformations occurring at the above-mentioned vacancy concentrations.

In the general case, the passage of a compression front through a crystal and the unloading wave that follows it causes the migration of vacancies towards the region of the crystal in which the wave was generated. Therefore, in the case of generation of several waves, vacancies do not have time to form clusters, and their localization in one place causes either amorphization of this region or the formation of a pore in it. In this regard, a comparative analysis of the structural changes occurring in the crystal during the generation of waves after a certain time interval was carried out. It turned out that the crystal in which the waves were generated is characterized by a decrease in the fraction of atoms belonging to the hcp phase. Thus, Fig. 92 shows the change in the corresponding number of atoms during the simulation with successive generations of five waves with an interval of 1,000 computational steps.

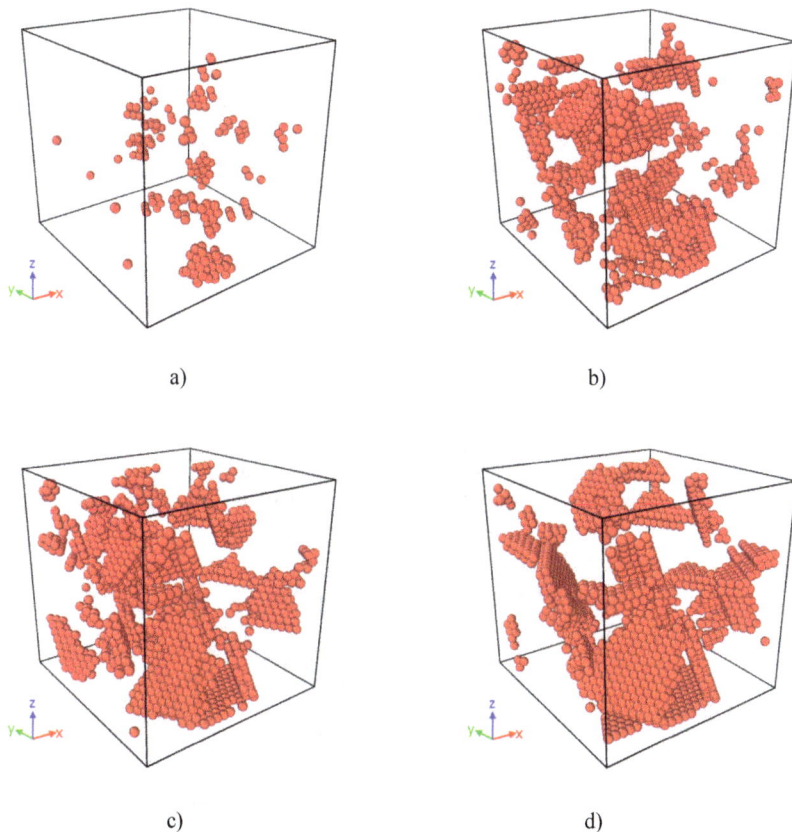

a)

b)

c)

d)

Figure 90: *Transformation of the defective structure of the computational cell after 1000 (a), 5000 (b), 10000 (c) and 15000 (d) computational steps. The vacancy concentration is n = 5%, the maintained temperature of the computational cell is T = 900 K. Atoms are shown, the local environment of which corresponds to the hcp lattice.*

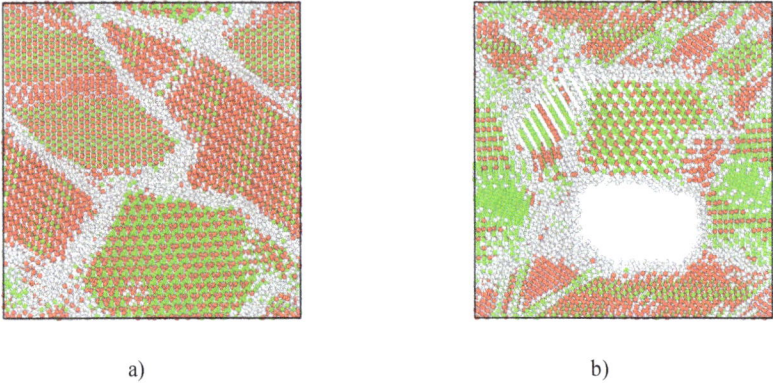

a)

b)

Figure 91: *Structure of the simulated crystal (XZY plane) after 10,000 computational steps: (a) n = 15%, T = 1200 K; (b) n = 20%, T = 300 K. Color visualization corresponds to the distribution of the local environment of atoms: green - fcc, red - hcp, white - not determined.*

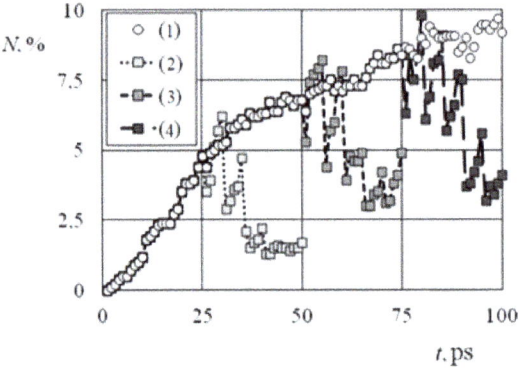

Figure 92: *Change in the content of N atoms of the computational cell belonging to the hcp phase during simulation: (1) – without wave generation; (2), (3) and (4) - when generating a wave after 5,000, 10,000, and 15,000 computational steps, respectively. The concentration of vacancies n = 5%, the maintained temperature of the computational cell is T = 900 K.*

An analysis of the dislocation structure of the computational cell showed that, after relaxation, the simulated system contains mainly vertex dislocations with the Burgers vector $\vec{b} = 1/6<110>$ and partial Shockley dislocations $\vec{b} = 1/6<112>$. The presence of these dislocations is due to stacking faults, including those forming a tetrahedron. After the passage of a series of shock waves, the total number of dislocation segments decreases significantly, and Shockley dislocations begin to have the greatest extent in the computational cell. For example, Fig. 93 shows the dislocation structures of the computational cell corresponding to the simulation results shown in Fig. 92 at the time t = 75 ps. At present, the total length of the lines of vertex dislocations for the simulated system was 1039.78 Å, and for Shockley dislocations, it was 1498.72 Å. If, however, five shock waves were generated in the computational cell by this time during the simulation, then the length of the dislocation lines decreased to 202.081 Å and 556.222 Å, respectively.

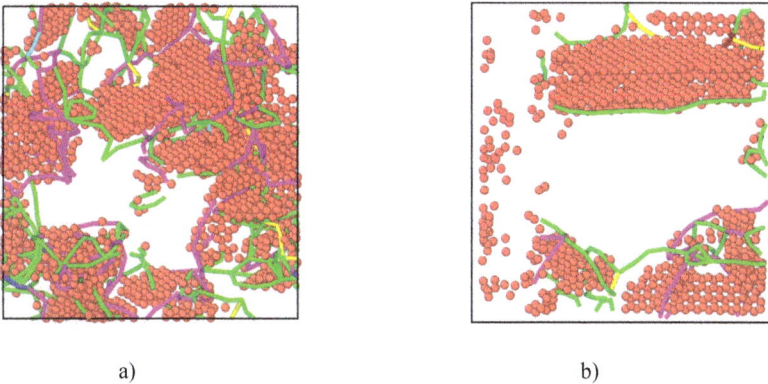

a)   b)

Figure 93: *Distribution of dislocation segments in the computational cell (XZY plane) after 75 ps simulation: (a) without generation of shock waves; (b) after the passage of five shock waves. The concentration of vacancies is n = 5%, the maintained temperature of the computational cell is T = 900 K. The color visualization of dislocation lines corresponds to the Burgers vector: green - 1/6<112>, violet - 1/6<110>, etc.*

When constructing images in Fig. 93, the Dislocation Extraction algorithm (DXA) was used, based on tessellation and Delaunay triangulation. This algorithm identifies the revealed dislocation lines and paints them in a certain color.

At the next stage, computational cells with a 15% vacancy concentration were studied. As mentioned above, in this case, the formation of a grain structure is typical. In this case, the excess free volume is distributed along the grain boundaries. Under the influence of waves, coarsening of grains is observed, and in this case, the distributed free volume begins to be localized in the region of wave generation (see Fig. 94).

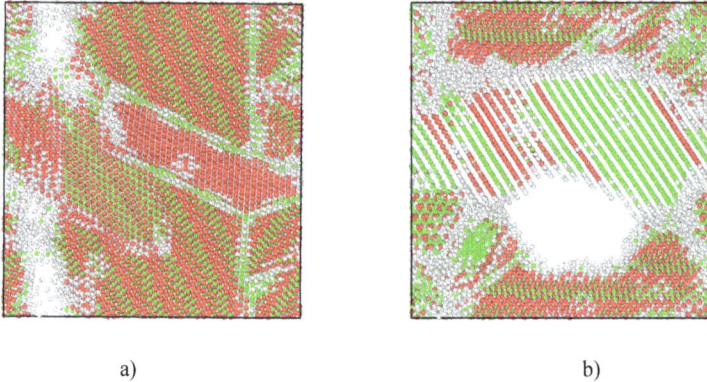

a)   b)

Figure 94: *Structure of the simulated crystal (XZY plane) after the passage of five shock waves generated after 3,000 computational steps: (a) n = 15%, T = 1200 K; (b) n = 20%, T = 300 K. Color visualization was performed similarly to Fig. 91.*

To quantify the dissolved free volume in grain boundaries, one can use the surface grid method based on Delaunay tetrahedrization. In this method, a geometric set of points is associated with a set of tetrahedra that fill the space between the points, provided that they are located at a certain distance (in our case, half the shortest distance between atoms). After that, the volume of the resulting shape is calculated.

Calculations using the method described above showed that, for example, after passing through the computational cell containing a 15% concentration of vacancies, five shock waves were generated with an interval of 3000 computational steps at temperatures of 600 K, 900 K, and 1200 K, the content of dissolved free volume decreases by 3.74%, 4.69%, and 6.73%, respectively.

The study has shown that post-cascade shock waves generated in a solid containing a depleted zone can have a significant effect on the processes of structural rearrangement of vacancy clusters. Under the influence of waves, the number of formed stacking fault tetrahedra decreases, and the fraction of atoms belonging to the hcp phase also decreases. In addition, the passage of waves along the emerging polycrystalline structure contributes to a decrease in the proportion of excess free volume dissolved at grain boundaries. As a rule, under the influence of waves, the free volume is localized in the form of nanopores.

6.2 The process of healing pores under the influence of shock waves

Structural imperfections of the crystalline structure of the body, and especially volume, play an important role in changing its physical and mechanical properties. So, for example, under high-intensity external influences (radiation, laser, etc.), intense pore formation occurs in a solid body, which significantly worsens the operational properties of the material. High porosity after sintering is the cause of corrosion propensity for products made by powder metallurgy. Therefore, the reduction of discontinuities is one of the most important tasks of modern materials science.

There are several technologies aimed at restoring the continuity of materials, including exposure to high temperatures or pressures. The choice of technologies used is largely determined by the position of the pores in the solid. So, for example, the use of only temperature exposure to healing pores that have access to the surface is ineffective, since in this case very high temperatures are required, leading to significant structural changes. Therefore, it is more efficient to use both of the above technologies. In addition, it should be noted that under high-intensity external action on a solid body, there is another factor that contributes to significant structural changes - shock waves. Previously, the authors carried out studies confirming the possibility of structural transformations of nanopores under the influence of such waves. In this case, a redistribution of the free volume, initially localized in the form of pores, is observed. The main mechanism of void healing in crystalline solids under external force is the emission of dislocation loops. In this section, the processes of dislocation nucleation on the surface of cylindrical pores are considered. Defects of this kind can be tracks that form after high-energy ions pass through the crystal, or, for example, when superheated closed inclusions of liquid (mother liquor) reach the surface. This category can also include, in fact, any extended defects that create a free surface in the volume of the metal, for example, fistulas that occur when gas is released during welding.

The purpose of the research, the results of which are presented in this section, is to establish the mechanisms of healing of cylindrical pores in an fcc crystal (for example, a gold crystal) under the influence of shock waves and to determine the parameters of the external action required for this.

To create a pore in the computational cell, a region was selected in the form of a cylinder, the axis of which was perpendicular to the (111) plane, after which the atoms belonging to it were removed. This was followed by a structural relaxation procedure implemented by running a computational algorithm, but with the velocities of atoms set to zero, and the resulting structure was used for further modeling.

The defect created in the computational cell is stable and an external action is required to activate its restructuring. Thus, to implement the healing of pores according to the dislocation mechanism, shear stresses are necessary, which contribute to the initiation of dislocation loops. The stress state in the computational cell, with a nonzero deviatoric component of the stress tensor, was achieved by shear deformation parallel to the (111) plane along the $[1\bar{1}0]$ direction. The geometry of the computational cell was maintained using a combination of rigid and periodic boundary conditions. Modeling showed that under such a given external action, partial Shockley dislocations in the form of dislocation loops begin to actively form (see Fig. 95).

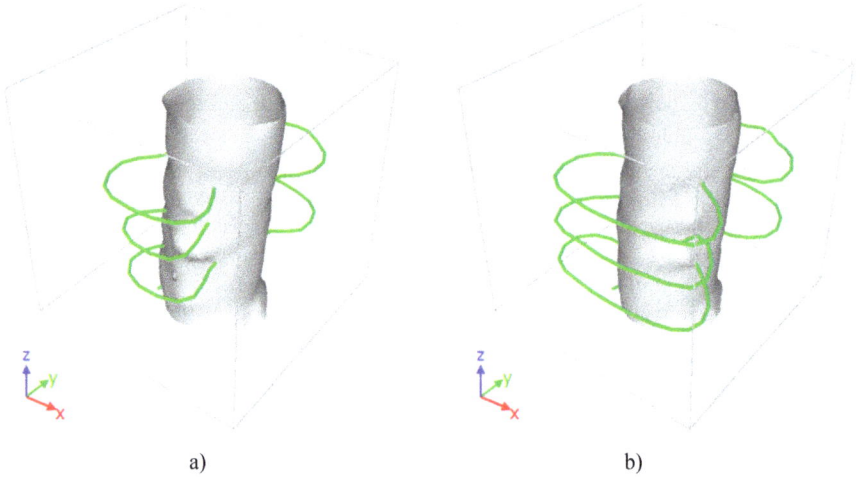

a)   b)

Figure 95: *Visualization of the development of the dislocation structure during the simulation after 1500 (a) and 5000 (b) calculation steps. Shear angle γ = 0.1 rad.*

Fig. 95 shows the surface formed by a cylindrical pore located in the computational cell and a set of dislocation segments. To identify dislocation lines in the simulated crystal structure and their subsequent visualization, a method based on tessellation and Delaunay triangulation was used. When constructing the pore image, the surface grid method based on Delaunay tetrahedrization was used.

The dimensions of the dislocation loops increase during the simulation until they reach values at which the resultant of all forces acting on the dislocation becomes equal to zero. In this case, as follows from the dependence shown in Fig. 96a, the total length of dislocation loops increases with an increase in the radius of the base of the created cylindrical nanopores, and, as a result, an increase in the area of the free surface in the computational cell, which is the source of heterogeneous nucleation of dislocations. The development of dislocation loops contributes to the drop in shear stresses in the calculation cell (see Fig. 96b).

Estimates show that the rate at which dislocation loops grow reaches ≈1600 m/s (for comparison, the velocity of longitudinal elastic waves in gold is 3240 m/s). But it should be taken into account that the results described above were obtained during simulation with constant removal of the thermal background by zeroing the velocities of atoms. In the case of modeling at a maintained temperature, the loop growth rate increases, which contributes to a faster relaxation of shear stresses.

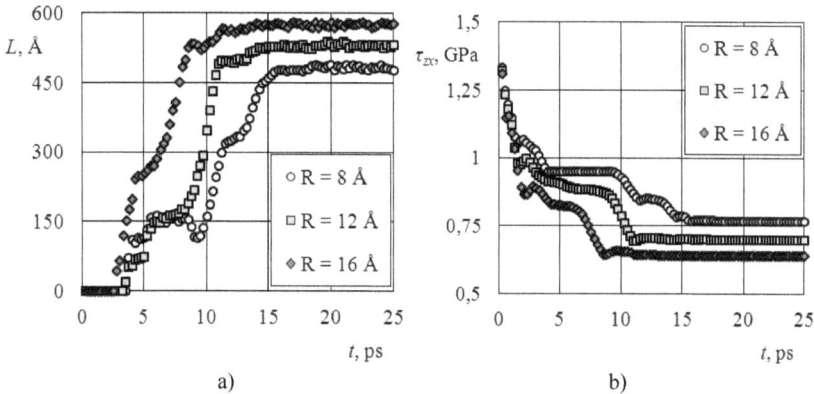

Figure 96: Change in the total length L of dislocation loops (a) and shear stresses in the computational cell (b) during the simulation at different radii of the base of cylindrical pores R. Shear angle $\gamma = 0.1$ rad.

To characterize the emerging dislocation structure, the dislocation density was calculated, which is equal to the ratio of the total length of dislocation segments to the volume of the computational cell filled with particles. The change in dislocation density at different shear angles is shown in Fig. 97a. The shear angle γ was set discretely with a step of 0.01 rad (in this case, the nucleation of dislocations began to be observed at $\gamma = 0.09$ rad). The calculations were carried out in two ways: with the removal of the thermal background and while maintaining a constant temperature. In the first case, the constructed dependence is close to linear, and in the second case, it has a flat section. For both cases, the calculated density values obtained during the simulation correspond to highly deformed unannealed samples of real metals. When modeling with a set temperature of the computational cell, the nucleation on the pore surface of many unstable "nuclei" of dislocation loops is observed, the mechanism of formation of which is of a thermal nature. These nuclei are pulled back to the surface when stable dislocation loops are formed, the number of which exceeds the number of loops in the simulation with the removal of the thermal background. The flat section of the curve is because, in this case, the relaxation of shear stresses is realized not by the development of a dislocation system, but in a different way, - by the formation of additional voids.

As the temperature increases, as a rule, the density of dislocations should decrease, which is due to the activation of the processes of sliding and climbing of dislocations, which contribute to their annihilation. In our case, the density also decreases (see Fig. 97b), but a different mechanism is realized. As the temperature increases, the process of amorphization of the structure begins in the region surrounding the cylindrical pore, and dislocation lines in it are not identified. A possible mechanism of near-surface amorphization is the reduced internal pressure in the crystal near the free surface. The higher value of the dislocation density at a smaller pore diameter is because in this case, as the temperature rises, the pore "collapses" and an additional dislocation structure is formed in its place.

Note that when plotting the graphs in Fig. 97, we took the average lengths of dislocation lines calculated over 500 simulation steps, since the loops change their size due to the thermal vibrations of atoms.

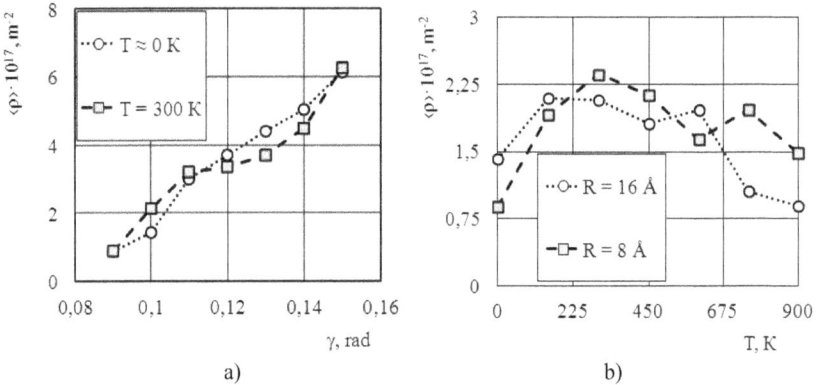

a)   b)

Figure 97: *Change in the average dislocation density ⟨ρ⟩ at different shear angles (R = 16 Å) (a) and different temperatures of the computational cell (γ = 0.1 rad) (b), calculated over a time interval of 12.5...15 ps.*

At the next stage of the study, a simulation was carried out with the generation of shock waves in the computational cell. The passage of the wavefront creates shear stresses sufficient, in particular, to initiate the process of sliding edge dislocations. Therefore, it is of interest to study the possible influence of shock waves on the process of structural changes occurring with the pore. In this case, we consider deformation with a shear angle at which the formation of dislocation loops does not occur, for example, γ = 0.07 rad (see Fig. 97a).

The simulation results showed that, under the influence of waves, dislocation loops begin to appear on the surface of the pore. Thus, Fig. 98 shows the color visualization of atoms, performed after structural analysis, which consists in identifying the local environment of particles using the Ackland-Jones method of angles and bonds, at a time of 5 ps after the passage of one shock wave. This visualization allows us to successfully detect packaging defects. As follows from the figure, for a smaller pore radius, the fraction of atoms with a local hcp environment exceeds the similar fraction for the computational cell containing a pore with a larger radius.

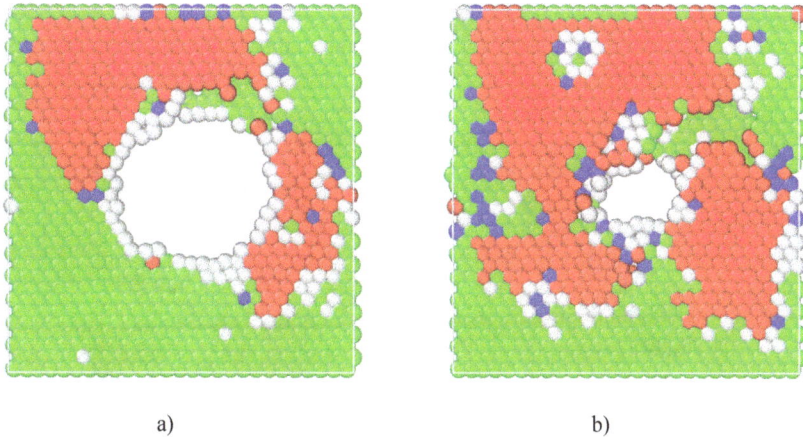

a)   b)

Figure 98: *Fragment of the (111) plane of the computational cell containing a pore with a base radius R = 16 Å (a) and 8 Å (b), 5 ps after the shock wave was generated. The temperature of the computational cell is T = 300 K. Color visualization corresponds to the distribution of the local environment of atoms: green - fcc, red - hcp, blue - bcc, white - undetermined.*

Fig. 99a shows the results of calculating the dislocation density for various radii of the base of a cylindrical pore. As follows from the figure, the dislocation density increases more significantly in the case of the smallest of the considered radii indicating the greatest ongoing structural changes, which consist in the dissolution of the pore under the influence of waves.

To characterize the process of structural transformations, the pore volume was calculated, which changes during the simulation. However, we believe that such a value as the specific volume of the "substance phase", which is the ratio of the volume occupied by the particles of the system to the total volume of the calculated cell, is more illustrative. Similar terminology is used, for example, in the physics of sintering ("substance phase", "emptiness phase") or the study of metallic glasses. The calculation results are shown in Fig. 99b. As follows from the presented dependences, the specific volume increases abruptly some time after the wave generation. These jumps are due to a change in the shape of the pore (compression of the pore by the compression front of the wave) and its subsequent recovery. The unloading wave formed later creates tangential stresses that promote the nucleation of dislocations. The subsequent increase in the specific volume of the "substance phase" indicates the activation of the pore healing process. As the temperature rises, the waves initiate the detachment of some of the vacancies, as a result of which the pore with the smallest simulated radius (R = 8 Å) loses its stability and partially or completely dissolves (the specific volume of the "substance phase" becomes equal to 1). Consequently, there is a delocalization of the free volume.

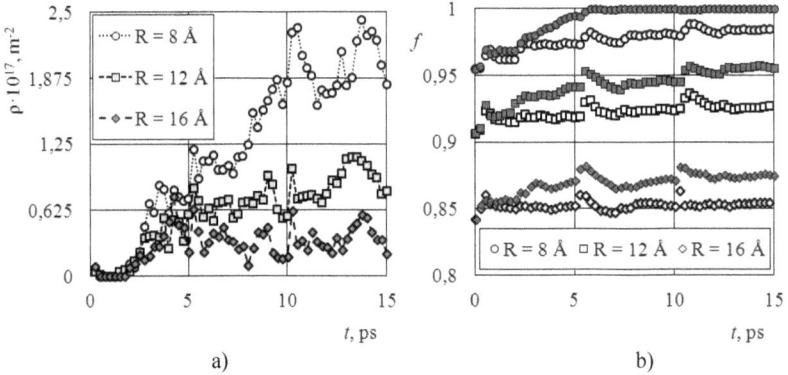

Figure 99: *Change in the dislocation density ρ (T = 300 K) (a) and the volume fraction of the "substance phase" f (light markers - T = 300 K, dark markers - T = 600 K) (b), during the passage of shock waves, generated in the computational cell with an interval of 5 ps. Shear angle γ = 0.07 rad.*

Thus, the study showed that shock waves can initiate the formation of dislocation loops on the surface of cylindrical pores, increasing the dislocation density in the crystal, which affects the strength properties of the metal. This process occurs at shear deformation, which is insufficient for the nucleation of dislocations. Under simulation conditions at a temperature of 600 K, the dissolution of the pore in the computational cell is observed. This indicates that shock waves, coupled with thermal action, are capable of initiating the process of pore healing.

Conclusion

The effects considered in the work affect only a small part of the possible manifestations of the nonlinearity of discrete media in the example of models of crystal lattices of various compositions and stoichiometry. The formation of localized structures shows the diversity of self-organization of such simple, at first glance, discrete structures.

The theoretical and experimental results of many works demonstrate the contribution of such objects to the macro properties of crystals, including thermal conductivity, heat capacity, hardness, plasticity, etc. The studies presented here using atomistic modeling methods expand the understanding of several effects of energy localization and related processes: solitons, discrete breathers, quasi-breathers, and shock waves. The possibility of structural and energy changes occurring in materials at the atomic and nanostructural levels, due to the propagation of soliton-like and shock waves initiated by intense external influences, is shown.

The main attention was paid to the study of solitary waves initiated by an external periodic action, discrete breathers, supersonic and shock waves in gratings based on fcc structures of various stoichiometric compositions. Thus, discrete breathers in monatomic crystals and biatomic crystals of composition AB and A_3B were studied. Note that in mathematical physics, a discrete breather is understood as strictly periodic, undamped in time oscillatory modes, but in real systems, where the presence of all kinds of perturbations is unavoidable, one should consider quasi-breathers that have a nonstrict periodicity of oscillations and a finite lifetime. The book uses both terms, essentially describing the same processes from different points of view. The statistical approach to the description of nonlinear localized modes made it possible to form a criterion for their destruction, thereby providing a theoretical tool for predicting their lifetime. Thus, knowing the formalized description of the localized mode and the entire structure, one can describe the process of energy dissipation in the crystal and estimate the time of the destruction of such modes. All objects considered from the position of quasi-breathers are destroyed at the moment when the standard deviation of frequencies exceeds the difference between the average frequency of the quasi-breather and the nearest boundary of the phonon spectrum of the crystal. In this case, vibrations are delocalized and energy is dissipated in the crystal in the form of low-amplitude thermal vibrations of the lattice.

For the first time, the mechanism of excitation of solitary waves in A_3B crystals under periodic external influences is disclosed, and the role of discrete breathers with a soft type of nonlinearity in the process of wave formation is shown. It has been established that such waves can propagate thousands of nanometers deep into a defect-free crystal without changing their shape and speed. The total amount of energy carried by the wave is determined by the number of rows of atoms involved in the oscillations, we can talk about tens and hundreds of eV. This mechanism of energy transport through the crystal, utilizing solitary waves, seems to be one of the most efficient, and the mechanism for generating such waves is relatively simple. Discrete breathers here act as a kind of accumulator of vibrational energy, which subsequently transforms into a solitary wave.

This reveals one of the mechanisms of nonlinear supratransmission in discrete structures with a forbidden phonon band.

The effect of shock waves on the defect structure of fcc crystals is considered. It is shown that in the process of relaxation of the simulated system containing a low concentration of vacancies, the defects are rearranged into stacking fault tetrahedra, and at a high concentration, the formation of a grain structure and pore formation are observed. Under the action of shock waves, the number of atoms belonging to the hcp phase and representing stacking faults in the simulated crystal decreases. In addition, by analyzing the dislocation structure of the system being modeled, the number, type, and total length of dislocation segments formed in the process of relaxation were estimated. It is shown that, under the action of shock waves, the total number of dislocation segments decreases, as a result of which segments of the type of partial Shockley dislocations begin to predominate. The study of the grain structure formed at a high concentration of vacancies showed that the excess free volume dissolves in grain boundaries. After the passage of shock waves, the fraction of the dissolved free volume decreases, and it is localized in the region of wave generation in the form of nanopores. A quantitative assessment of the decrease in the dissolved free volume at various temperatures has been made. In addition, when considering the pores of a cylindrical handicap, the dependences of the average dislocation density on the shear angle and temperature of the computational cell were established, and the loop growth rate was estimated. The generated shock waves create additional shear stresses that contribute to the formation of dislocation loops; therefore, in this case, dislocations are observed even at low shear strain. If during the simulation the thermal effect increases, then the pore collapses.

The obtained estimated numerical values of the energy, velocity, lifetime, and other characteristics of the objects under consideration give a general idea of such lattice states, which can simplify their study in field experiments. The results can be generalized to a wider range of crystals and alloys of various compositions and structures. Some of the materials considered are the basis of superalloys. It is possible to apply the results obtained in educational activities, so a number of molecular dynamics models have been converted into 3D models for virtual and mixed reality systems.

Acknowledgments

The team of authors is deeply grateful to colleagues Dmitriev S.V., Medvedev N.N., Korznikova E.A., Lobzenko I.P., Lutsenko I.S., Cherednichenko A.I., and Eremin A.M. for the discussion and preparation of scientific research materials.

References

[1] ABINIT https://www.abinit.org/

[2] Ackland G.J., Jones A.P. Applications of local crystal structure measures in experiment and simulation // Physical Review B. 2006. Vol.73. No.5. 054104 (7 p) https://doi.org/10.1103/PhysRevB.73.054104

[3] Andersen H.C. Molecular dynamics simulations at constant pressure and/or temperature // The Journal of Chemical Physics. 1980. Vol.72(4), pp. 2384-2393 https://doi.org/10.1063/1.439486

[4] Cherne F.J., Baskes M.I. Properties of liquid nickel: A critical comparison of EAM and MEAM calculations // Physical Review B. 2001. Vol. 65(2). 024209(9) https://doi.org/10.1103/PhysRevB.65.024209

[5] Chudinov V.G., Cotterill R.M.J., Andreev V.V. Kinetics of the diffuse processes within a cascade region in the sub-threshold stages of F.C.C. and H.C.P. Metals // Physica Status Solidi (a). 1990. Vol.122(1), pp. 111-120 https://doi.org/10.1002/pssa.2211220110

[6] Dmitriev, S.V., Medvedev, N.N., Chetverikov, A.P., Zhou, K. and Velarde, M.G. (2017), Highly Enhanced Transport by Supersonic N-Crowdions. Phys. Status Solidi RRL, 11: n/a, 1700298. doi:10.1002/pssr.201700298 https://doi.org/10.1002/pssr.201700298

[7] Dudarev S.L. Density functional theory models for radiation damage // Annual Review of Materials Research. 2013. V.43. P. 35 - 61 https://doi.org/10.1146/annurev-matsci-071312-121626

[8] Edelsbrunner H., Mücke E. Three-dimensional alpha shapes // ACM Transactions on Graphics. 1994. Vol.13. №1. Pp.43-72 https://doi.org/10.1145/174462.156635

[9] Inogamov, N.A., Zhakhovskii, V.V., Khokhlov, V.A. et al. Jetp Lett. (2011) 93: 226. https://doi.org/10.1134/S0021364011040096 https://doi.org/10.1134/S0021364011040096

[10] J.F.R. Archilla, S.M.M. Coelho, F.D. Auret, C. Nyamhere, V.I. Dubinko, V. Hizhnyakov Experimental observation of moving intrinsic localized modes in germanium // arXiv:1311.4269 [cond-mat.mtrl-sci]

[11] Johnson R.A. Alloy models with the embedded-atom method // Physical Review B. 1989. V.39. №17. P. 12554 - 12559 https://doi.org/10.1103/PhysRevB.39.12554

[12] Johnson R.A. Analytic nearest-neighbor model for fcc metals // Physical Review B. 1988. V.37. №8. P.3924-3931 https://doi.org/10.1103/PhysRevB.37.3924

[13] LAMMPS Molecular Dynamics Simulator // [Electronic resource] Mode of access: http://lammps.sandia.gov/

[14] Manley M.E., Sievers A.J., Lynn J.W., Kiselev S.A., Agladze N.I., Chen Y., Llobet

A., Alatas A. Intrinsic Localized Modes Observed in the High Temperature Vibrational Spectrum of NaI // Phys. Rev. B. - 2009. - V.79. 134304 https://doi.org/10.1103/PhysRevB.79.134304

[15] Nordlund K., Keinonen J., Ghaly M., Averback R.S. Coherent displacement of atoms during ion irradiation // Nature. 1999. V.398. P.49-51 https://doi.org/10.1038/17983

[16] Pushkarov D.I. Quantum theory crowdions at low temperatures // Journal of Experimental and Theoretical Physics. 64. 1973. p. 634

[17] Quantum Espresso http://www.quantum-espresso.org/

[18] RasMol and OpenRasMol. Molecular graphics visualisation tool // [Electronic resource]. Mode of access: http://www.rasmol.org/

[19] Sheng H.W., Kramer M.J., Cadien A., Fujita T., Chen M.W. Highly optimized embedded-atom-method potentials for fourteen FCC metals // Physical Review B. 2011. V.83. №13. P.134118-134128 https://doi.org/10.1103/PhysRevB.83.134118

[20] Stukowski A. Computational analysis methods in atomistic modeling of crystals // The Journal of The Minerals, Metals & Materials Society. 2014. Vol.66. No.3. Pp.399-407 https://doi.org/10.1007/s11837-013-0827-5

[21] Stukowski A. Visualization and analysis of atomistic simulation data with OVITO - the Open Visualization Tool // Modelling and Simulation Materials Science and Engineering. 2010. Vol.18. 015012 (7 pp) https://doi.org/10.1088/0965-0393/18/1/015012

[22] Stukowski A., Albe K. Extracting dislocations and non-dislocation crystal defects from atomistic simulation data // Modelling and Simulation in Materials Science and Engineering. 2010. Vol.18. No8. 085001 (13 pp) https://doi.org/10.1088/0965-0393/18/8/085001

[23] Stukowski A., Bulatov V.V., Arsenlis A. Automated identification and indexing of dislocations in crystal interfaces // Modelling and Simulation in Materials Science and Engineering. 2012. Vol.20. No.8. 085007 (16 pp) https://doi.org/10.1088/0965-0393/20/8/085007

[24] Taubin G. A signal processing approach to fair surface design // Proceedings of the 22nd annual conference on Computer graphics and interactive techniques « SIGGRAPH '95». - ACM New York, NY, USA, 1995. Pp.351-358 https://doi.org/10.1145/218380.218473

[25] Welcome to the EAM Alloy Potential Generator // [Electronic resource]. Mode of access: http://atomistics.osu.edu/eam-potential-generator/index.php

[26] XMD - Molecular Dynamics for Metals and Ceramics // [Electronic resource]. Mode of access: http://xmd.sourceforge.net/about.html

[27] Zhukov V.P., Boldin A.A. Elastic-wave generation in the evolution of displacement

peaks // Atomic Energy. 1987. V. 68. P. 884 - 889
https://doi.org/10.1007/BF01126098

[28] Aksenov M.S., Poletaev G.M., Rakitin R.Yu., Krasnov V.Yu., Starostenkov M.D.
Stability of vacancy clusters in fcc metals // Fundamental problems of modern
materials science. 2005. V.2. No. 4. pp. 24 - 31

[29] Alalykin A.S., Krylov P.N., Fedotova I.V., Fedotov A.B. Influence of treatment with
low-energy Ar ions on the characteristics of the working and back sides of the single-
crystal GaAs substrate // FTPP, 2003, vol. 37, no. 4, pp. 465-468
https://doi.org/10.1134/1.1568466

[30] Borovitskaya I.V., Dedyurin A.I., Ivanov L.I., Krokhin O.N., Nikulin V.Ya., Petrov
V.S., Tikhomirov A.A. Changes in the bulk properties of vanadium under the
influence of high-temperature dense pulsed deuterium plasma // Perspektivnye
materialy. 2004. No. 2. S. 44-48

[31] Grigoriev I.S., Meilikhov E.Z. Physical quantities: reference book. - Moscow:
Energoatomizdat, 1991. - 1232 p.

[32] Groza A.A., Litovchenko P.G., Nikolaeva L.G., Starchik M.I., Khivrich V.I.,
Shmatko G.G. Long-range effects in silicon single crystals under irradiation with
protons and alpha particles http://jnpae.kinr.kiev.ua/11.1/Abstracts/jnpae-abstract-ru-
2010-11-0066-Groza.pdf

[33] Demidov E.S., Gromoglasova A.B., Karzanov V.V. Long-range effect of ion
irradiation, chemical etching and mechanical grinding on the relaxation of a solid
solution of iron in gallium phosphide // FTPP, 2000, vol. 34, no. 9, pp. 1025-1029
https://doi.org/10.1134/1.1309397

[34] Dmitriev S.V., Korznikova E.A., Baimova Yu.A., Velarde M.G. Discrete breathers
in crystals // UFN 186 471-488 (2016) https://doi.org/10.3367/UFNr.2016.02.037729

[35] Zakharov P.V. Module for calculating the density of phonon states of model crystals
by the method of molecular dynamics // Rospatent. Certificate of state registration of
the computer program No. 2015614597 dated April 21, 2015

[36] Kistanov A.A., Semenov A.S., Dmitriev S.V. Properties of moving discrete
breathers in a monoatomic two-dimensional crystal. // ZhETF. 2014. V.146. S. 869
https://doi.org/10.1134/S1063776114100045

[37] Markidonov A.V. On the possibility of creating capillary structures in a crystal by
dividing latent tracks by shock waves (computer simulation) // Bulletin of the
Kuzbass State Technical University. 2014. No. 1 (101). pp.99-103

[38] Markidonov A.V., Zakharov P.V., Starostenkov M.D., Medvedev N.N. Mechanisms
of cooperative behavior of atoms in crystals. - Novokuznetsk: branch of KuzGTU in
Novokuznetsk, 2016. - 219 p.

[39] Markidonov A.V., Poletaev G.M., Starostenkov M.D. The role of high-speed

cooperative atomic displacements in superdeep penetration of a substance during radiation irradiation of a material // Fundamental problems of modern materials science. 2012. V.9. No. 2. pp. 201-208

[40] Markidonov A.V., Starostenkov M.D. Coalescence of vacancy nanopores in an fcc lattice under the influence of post-cascade shock waves. Bulletin of the Voronezh State University. Series "Physics. Maths". 2016. No. 1. pp. 14-23

[41] Markidonov A.V., Starostenkov M.D. On the possibility of homogeneous generation of a pore in the grain-boundary region under the impact of post-cascade shock waves. Voprosy Atomnoi Nauki i Tekhniki. Series: "Mathematical modeling of physical processes". 2016. No. 3, pp. 37-46

[42] Markidonov A.V., Starostenkov M.D. Radiation-dynamic processes in fcc crystals accompanied by high-speed mass transfer. - Kemerovo: Kuzbassvuzizdat, 2014 - 191p.

[43] Markidonov A.V., Starostenkov M.D., Barchuk A.A. Interaction of moving crowdion complexes with point defects in an fcc crystal // Fundamental problems of modern materials science. 2012. V.9. No. 1. pp. 108-113

[44] Markidonov A.V., Starostenkov M.D., Barchuk A.A., Bovkush S.V. Energy dissipation of moving crowdions at low-angle tilt grain boundaries in aluminum // Fundamental Problems of Modern Materials Science. 2011. V.8. No. 4. pp.99-103

[45] Markidonov A.V., Starostenkov M.D., Barchuk A.A., Bovkush S.V. Peculiarities of crowdion dynamics in fcc crystals under various force impacts // Chemical Physics and Mezoscopy. 2012. V.14. No. 1. pp.46-54

[46] Markidonov A.V., Starostenkov M.D., Barchuk A.A., Medvedev N.N. Peculiarities of the dynamics of crowdions and their complexes in a deformed fcc crystal // Fundamental problems of modern materials science. 2011. V.8. Number 3. pp.83-88

[47] Markidonov A.V., Starostenkov M.D., Grishunin V.A., Baschenko L.P. Atomic mechanisms of migration of the boundary of torsion grains (110) under the influence of post-cascade shock waves on the example of nickel // Fundamental problems of modern materials science. 2017. V.14. No. 4. pp.528-534

[48] Markidonov A.V., Starostenkov M.D., Zakharov P.V. Growth of small vacancy clusters initiated by post-cascade shock waves // Letters on Materials. 2012. V.2. Issue 2. pp. 111-114 https://doi.org/10.22226/2410-3535-2012-2-111-114

[49] Markidonov A.V., Starostenkov M.D., Zakharov P.V., Zhiliang Zhang. The role of post-cascade shock waves in low-temperature activation of self-diffusion // Fundamental problems of modern materials science. 2014. V.11. Number 3. pp.346-353

[50] Markidonov A.V., Starostenkov M.D., Zakharov P.V., Obidina O.V. Pore formation in an fcc crystal under the influence of post-cascade shock waves // Fundamental

problems of modern materials science. 2015. V.12. No. 2. pp.231-240.

[51] Markidonov A.V., Starostenkov M.D., Obidina O.V. Aggregation of vacancies initiated by post-cascade shock waves // Fundamental problems of modern materials science. 2012. V.9. No. 4. pp.548-555

[52] Markidonov A.V., Starostenkov M.D., Pavlovskaya E.P. Influence of post-cascade shock waves on the processes of enlargement of vacancy pores // Fundamental problems of modern materials science. 2012. V.9. No. 4/2. pp.694-702

[53] Markidonov A.V., Starostenkov M.D., Pavlovskaya E.P. Influence of post-cascade shock waves on structural transformations of vacancy pores // Chemical Physics and Mezoscopy. 2013. V. 15. No. 3. pp.370-377

[54] Markidonov A.V., Starostenkov M.D., Pavlovskaya E.P., Yashin A.V., Medvedev N.N., Zakharov P.V. Structural transformation of vacancy pores in a deformed crystal under the influence of shock waves // Fundamental problems of modern materials science 2013. V.10. No. 4. pp.563-572

[55] Lutsenko I.S., Starostenkov M.D., Zakharov P.V., Dmitriev S.V., Korznikova E.A. Stability of supratransmission waves in a crystal of a3b stoichiometry upon interaction with single dislocation // Journal of Physics: Conference Series. Cep. "International Conference PhysicA.SPb/2021" 2021. P. 012079. https://doi.org/10.1088/1742-6596/2103/1/012079

[56] Markidonov A.V., Starostenkov M.D., Pavlovskaya E.P., Yashin A.V., Medvedev N.N., Zakharov P.V., Sitnikov A.A. Splitting of a vacancy pore in the grain-boundary region by a post-cascade shock wave // Fundamental problems of modern materials science. 2013. V.10. Number 3. pp.443-453.

[57] Zakharov P.V., Eremin A.M., Safronova S.A., Markidonov A.V., Starostenkov M.D. Dislocation structure of dicotyledonous bimetals under external harmonic influence: molecular dynamics modeling // IOP Conference Series: Materials Science and Engineering. Krasnoyarsk Science and Technology City Hall of the Russian Union of Scientific and Engineering Associations. 2020. P. 22010. https://doi.org/10.1088/1757-899X/862/2/022010

[58] Zakharov P.V., Lutsenko I.S., Markidonov A.V., Cherednichenko A.I. Modeling the interaction of pbse nanoparticleS // Journal of Physics: Conference Series. II International Scientific Conference on Metrological Support of Innovative Technologies (ICMSIT II-2021). Krasnoyarsk, 2021. P. 22006. https://doi.org/10.1088/1742-6596/1889/2/022006

[59] Markidonov A.V., Starostenkov M.D., Pavlovskaya E.P., Yashin A.V., Poletaev G.M. Low-temperature dissolution of a pore near the surface of a crystal under the influence of shock waves // Fundamental problems of modern materials science. 2013. V.10. No. 2. pp.254-261

[60] Shepelev I.A., Korznikova E.A., Dmitriev S.V., Zakharov P.V. Energy exchange of

a m-soliton cluster in a 2d morse lattice // Progress in Biomedical Optics and Imaging - Proceedings of SPIE. 7, Computations and Data Analysis: from Nanoscale Tools to Brain Functions. Cep. "Saratov Fall Meeting 2019 - Computations and Data Analysis: from Nanoscale Tools to Brain Functions" 2020. P. 114590Y. https://doi.org/10.1117/12.2565759

[61] Zakharov P.V., Eremin A.M., Starostenkov M.D., Lutsenko I.S., Korznikova E.A., Dmitriev S.V. Excitation of soliton-type waves in crystals of the a3b stoichiometry // Physics of the Solid State. 2019. V. 61. № 11. pp. 2160-2166. https://doi.org/10.1134/S1063783419110416

[62] Zakharov P.V., Korznikova E.A., Dmitriev S.V., Ekomasov E.G., Zhou K. Surface discrete breathers in pt3al intermetallic alloy // Surface Science. 2019. T. 679. P. 1-5. https://doi.org/10.1016/j.susc.2018.08.011

[63] Markidonov A.V., Starostenkov M.D., Poletaev G.M. Transformation of nanopores in gold under conditions of thermal activation and exposure to sound and shock waves // Izvestiya RAN. Physical series. 2015. V.79. No. 9. C.1233-1237 https://doi.org/10.3103/S1062873815090130

[64] Markidonov A.V., Starostenkov M.D., Potekaev A.I., Medvedev N.N., Neverova T.I., Barchuk A.A. Behavior of crowdions and their complexes in the weakly stable state of materials. Izvestiya VUZov. Physics. 2011. V.54. No. 11. pp.61-67 https://doi.org/10.1007/s11182-012-9737-1

[65] Markidonov A.V., Starostenkov M.D., Smirnova M.V. The process of self-diffusion in an fcc crystal caused by the passage of a shock wave. Izvestiya VUZov. Physics. 2015. V.58. No. 6. S.80-84 https://doi.org/10.1007/s11182-015-0576-8

[66] Markidonov A.V., Starostenkov M.D., Soskov A.A., Poletaev G.M. Study of structural transformations of cylindrical nanopores in gold by the method of molecular dynamics under conditions of thermal activation and exposure to sound and shock waves // Solid State Physics. 2015. V.57. No. 8, pp. 1521-1524 https://doi.org/10.1134/S106378341508017X

[67] Markidonov A.V., Starostenkov M.D., Tikhonova T.A., Barchuk A.A. Mechanisms of transformation of crowdion complexes during the passage of a longitudinal wave // Nonlinear World. 2011. V.9. No. 12, pp. 826-835

[68] Markidonov A.V., Starostenkov M.D., Yashin A.V. Structural transformations of a vacancy pore during radiation irradiation of a material // Fundamental problems of modern materials science. 2013. V.10. No. 1. pp.12-21

[69] Markidonov A.V., Starostenkov M.D., Yashin A.V., Zakharov P.V. Study of structural transformations of cylindrical pores by the method of molecular dynamics // Fundamental problems of modern materials science. 2014. V.11. No. 2. pp. 163-173

[70] Markidonov A.V., Tikhonova T.A., Nurkenova B.D., Poletaev G.M., Starostenkov M.D. Influence of longitudinal waves on complexes of point defects in an fcc crystal

// Proceedings of the Altai State University. 2010. No. 1/2(65). pp.175-178

[71] Markidonov A.V., Yashin A.V., Chaplygina A.A., Sinitsa N.V. Modeling of shock wave propagation in nanoobjects by the method of molecular dynamics (WAVE). Certificate of the state. registration of the computer program No. 2013661857, 17 Dec 2013

[72] Markidonov A.V., Starostenkov M.D., Zakharov P.V., Lubyanoy D.A., Lipunov V.N. Emission of dislocation loops by nanopores in an fcc crystal under the influence of post-cascade shock waves during shear deformation // Journal of Experimental and Theoretical Physics. 2019. V.156. No. 6(12). C.1078-1083 https://doi.org/10.1134/S106377611911013X

[73] Martynenko Yu.V., Moskovkin P.G. Long-range effect and energy transfer in solids during ion bombardment // Letters to the journal of Technical Physics. 1996. V.22. Issue 17. pp.54-58

[74] Zakharov P.V., Eremin A.M., Starostenkov M.D., Medvedev N.N., Dmitriev S.V. Simulation of the interaction between discrete breathers of various types in a pt3al crystal nanofiber // Journal of Experimental and Theoretical Physics. 2015. V. 121. № 2. pp. 217-221. https://doi.org/10.1134/S1063776115080154

[75] Medvedev N.N., Zakharov P.V. Modeling by the method of molecular dynamics of a two-dimensional crystal lattice of A3B stoichiometry with the possibility of fixing discrete breathers (DKR_A3B_DB). Certificate of state registration of the computer program No. 2010614584, dated 29 July 2010

[76] Medvedev N.N., Starostenkov M.D., Markidonov A.V., Zakharov P.V. Focusing and crowdion collisions of Cu atoms in a three-dimensional model of an ordered CuAu alloy with an L11 superstructure // Perspektivnye materialy. 2011. Special issue. No. 12, pp. 321-326

[77] Medvedev N.N., Starostenkov M.D., Zakharov P.V., Pozidaeva O.V. Localized oscillating modes in two-dimensional model of regulated pt3al alloy // Technical Physics Letters. 2011. V. 37. № 2. pp. 98-101. https://doi.org/10.1134/S1063785011020106

[78] Zakharov P.V., Eremin A.M., Starostenkov M.D., Dmitriev S.V., Korznikova E.A. Stationary quasi-breathers in monatomic fcc metals // Journal of Experimental and Theoretical Physics. 2017. V. 125. № 5 pp. 913-919. https://doi.org/10.1134/S1063776117100181

[79] Mirzoev F., Shelepin L.A. Solitary concentration waves of point defects under pulsed laser action // Letters to ZhTF, 1999, V.25, no. 16, pp. 90-94 https://doi.org/10.1134/1.1262594

[80] V.D. Natsik and S.N. Smirnov, Crowdions in the theory of elasticity, Crystallogr. 2009. V. 54. 6. C. 1034-1042 https://doi.org/10.1134/S1063774509060133

[81] Ovchinnikov V.V. Radiation dynamic effects. Possibilities of formation of unique structural states and properties of condensed media // Uspekhi Fizicheskikh Nauk. 2008. V.178. No. 9. pp.991-1001

[82] Pavlov P.V., Semin Yu.A., Skupov V.D., Tetelbaum D.I. Influence of elastic waves arising during ion bombardment on the structural perfection of semiconductor crystals // FTP, 1986, vol. 20, no. 3, pp. 503-507

[83] Zakharov P.V., Starostenkov M.D., Dmitriev S.V. Discrete breathers in biatomic crystals of ab and a 3 b composition // Bulletin of the Russian Academy of Sciences: Physics. 2017. V. 81. № 11. pp. 1322-1326. https://doi.org/10.3103/S1062873817110211

[84] Poletaev G.M. Atomic mechanisms of structural-energy transformations in the volume of crystals and near the boundaries of tilt grains in fcc metals / Dissertation of Physical and Mathematical Sciences. - Barnaul. 2008. - 356p.

[85] Poletaev G.M. Modeling by the method of molecular dynamics of structural and energy transformations in two-dimensional metals and alloys (MD2). Certificate of state registration of computer programs No. 2008610486, 25 Jan 2008

[86] Poletaev G.M. Modeling of structural-energy transformations in three-dimensional fcc metals (MD3) by the method of molecular dynamics. Certificate of state. Registration of computer programs No. 2008610487, 25 Jan 2008

[87] Psakhie S.G., Zolnikov K.P., Kadyrov R.I., Rudenskii G.E., Sharkeev Yu.P., Kuznetsov V.M. On the possibility of the formation of soliton-like pulses during ion implantation // Letters to ZhTF, 1999, vol. 25, no. 6, p. 7 https://doi.org/10.1134/1.1262425

[88] Serov I.N., Margolin V.I., Zhabrev V.A., Tupik V.A., Fantikov V.S. Long-range effects in micro and nanoscale structures // Engineering Physics, 2005, no. 1, pp. 50-67

[89] Starostenkov M.D., Markidonov A.V., Medvedev N.N., Tikhonova T.A. Simulation of mass transfer in the form of rows of vacancies and interstitial atoms on the example of a two-dimensional crystal // Bulletin of the Samara State Technical University. Series "Physical and Mathematical Sciences". 2010. Issue 1(20). pp.249-252

[90] Starostenkov M.D., Markidonov A.V., Pavlovskaya E.P. Structural transformations of vacancy pores under the influence of shock waves // Bulletin of the Tambov University. Series "Natural and technical sciences". 2013. V.18. Issue 4. Part 2. C.1743-1744

[91] Starostenkov M.D., Markidonov A.V., Tikhonova T.A., Medvedev N.N. High-speed mass transfer in a two-dimensional nickel crystal in the presence of dislocation loops of various local densities. Izvestiya VUZov. Ferrous metallurgy. 2009. No. 6. pp.57-60

[92] Starostenkov M.D., Markidonov A.V., Tikhonova T.A., Potekaev A.I., Kulagina V.V. High-speed mass transfer in crystalline aluminum containing chains of vacancies and interstitial atoms // Proceedings of universities. Physics. 2009, Vol. 52. No. 9/2. pp.139-145

[93] Zakharov P.V., Dmitriev S.V., Starostenkov M.D. Dynamics of discrete breathers in the pt3al crystal // Key Engineering Materials. 2016. V. 685. pp. 65-69. https://doi.org/10.4028/www.scientific.net/KEM.685.65

[94] Starostenkov M.D., Markidonov A.V., Tikhonova T.A., Potekaev A.I., Kulagina V.V. High-speed mass transfer in fcc metals containing chains of vacancies and interstitial atoms. Izvestiya VUZov. Physics. 2011. V.54. Number 3. pp.42-46 https://doi.org/10.1007/s11182-011-9616-1

[95] Starostenkov M.D., Potekaev A.A., Markidonov A.V., Kulagina V.V., Grinkevich L.S. Dynamics of edge dislocations in the weakly stable state of the fcc system under irradiation with high-energy particles. Physics. 2016. V.59. No. 9. pp.105-112 https://doi.org/10.1007/s11182-017-0929-6

[96] Yankovaskaya U.I., Korznikova E.A., Korpusova S.D., Zakharov P.V. Mechanical Properties of the Pt-CNT Composite under Uniaxial Deformation: Tension and Compression. Materials. 2023; 16(11):4140. https://doi.org/10.3390/ma16114140 https://doi.org/10.3390/ma16114140

[97] Zakharov P.V., Eremin A.M., Starostenkov M.D., Korznikova E.A., Dmitriev S.V. Excitation of gap discrete breathers in an a3b crystal with a flux of particles // Physics of the Solid State. 2017. V. 59. № 2. pp. 223-228. https://doi.org/10.1134/S1063783417020342

[98] Physical materials science: a textbook for universities. Ed. B.A. Kalina. - M.: Mifi, 2007. - ISBN 978-5-7262-0821-3

www.ingramcontent.com/pod-product-compliance
Lightning Source LLC
Chambersburg PA
CBHW071709210326
41597CB00017B/2400